T0329729

3DTV

3DTV
Processing and Transmission of 3D Video Signals

Anil Fernando

CVSSP, University of Surrey, UK

Stewart T. Worrall

Ericsson Television, UK

Erhan Ekmekcioğlu

CVSSP, University of Surrey, UK

This edition first published 2013

Registered office
John Wiley & Sons Ltd, The Atrium, Southern Gate, Chichester, West Sussex, PO19 8SQ, United Kingdom

For details of our global editorial offices, for customer services and for information about how to apply for permission to reuse the copyright material in this book please see our website at www.wiley.com.

Library of Congress Cataloging-in-Publication Data

Ekmekcioğlu, Erhan.
3DTV : processing and transmission of 3D video signals / Dr Erhan Ekmekcioğlu, Dr Anil Fernando, Dr Stewart Worrall.
1 online resource.
Includes bibliographical references and index.
Description based on print version record and CIP data provided by publisher; resource not viewed.
ISBN 978-1-118-70573-5 (Adobe PDF) – ISBN 978-1-118-70648-0 (ePub)
ISBN 978-1-118-70683-1 (MobiPocket) – ISBN 978-1-119-99732-0 (cloth)
1. 3-D television. I. Fernando, Anil. II. Worrall, Stewart. III. Title.
TK6658
621.388 – dc23

2013021021

A catalogue record for this book is available from the British Library.

ISBN: 978-1-119-99732-0

Set in 10/12pt Palatino by Laserwords Private Limited, Chennai, India
Printed and bound in Singapore by Markono Print Media Pte Ltd

1 2013

Contents

Preface

For more than 100 years, the popularity of 3D images has waxed and waned. Initial interest in the more immersive experience that 3D video can provide has often been confounded by expensive production costs and headache-inducing problems in visual quality. During the last decade, 3D video has experienced a revival that has proved more enduring than previous episodes. Despite the sceptics, movies continue to be produced in 3D, and many new televisions sold are capable of showing 3D video. This means that there is a large amount of 3D content available, and increasing numbers of 3D televisions in the home. All of this should create the best platform yet for 3DTV to become a success, and has led to broadcasters putting dedicated 3D channels into their multiplexes.

As one might expect, 3DTV introduces a number of new challenges compared to existing television production and broadcast scenarios. This book is intended to provide an introduction to the key concepts associated with 3DTV. It focuses on issues on the content creation side, delivery to the end-user, and how we measure the quality of 3D video. It also investigates display technologies. These are probably the most challenging areas associated with 3DTV.

Content creation represents a significant challenge, as correctly setting up multiple cameras is challenging. This is particularly true for live broadcasts. If stereoscopic viewpoints are incorrectly configured, viewers will be subjected to significant visual discomfort. If live broadcast is not required, then post-processing can be used to adjust viewpoints, and refine the visual experience.

Delivery to the end-user currently takes place by packing two stereoscopic views within a single frame. This is a relatively inexpensive method of providing 3DTV, as it is compatible with existing broadcast systems, and does not require significant additional bandwidth to be used. However, to deliver a truly immersive experience for more advanced displays, more than two views will need to be delivered to the end-user. This means that new coding techniques are required to reduce the bandwidth required by the multiple views. This book examines some of the compression approaches that may be used to reduce bandwidth requirements for multiple view transmission.

Displays play a very important role in the 3DTV chain. Consumer equipment should be affordable, should provide a reasonable quality viewing experience, and should not give end-users headaches. We examine some different display types within the book, looking at some of their advantages and disadvantages.

Finally, we examine issues associated with measuring the quality of video. Numerical metrics have often been used to measure the quality of video. These metrics are known to be far from perfect for 2D video. For 3D video it would seem to be even more important to think about how to measure quality. Chapter 6 in this volume is devoted to what affects 3D visual quality, and how it may be measured using numerical, objective approaches.

Acknowledgements

We would like to thank the following academics and researchers, whose valuable research contributions at the University of Surrey, have helped inform much of the writing contained in the book: Professor Ahmet Kondoz, Dr Lasith Yasakethu, Dr Varuna De Silva, Dr Safak Dogan, Dr Thushara Hewage, Dr Hemantha Araachchi, Dr Gokce Nur, Dr Omar Abdul-Hameed, Dr Marta Mrak.

Parts of the work presented throughout this book has been conducted within the European collaborative research projects DIOMEDES (Distribution of Multi-view Entertainment using Content-Aware Delivery Systems), MUSCADE (Multimedia Scalable 3D for Europe) and ROMEO (Remote-Collaborative Real-time Multimedia Experience over the Future Internet). We would like to thank the researchers who participated in these collaborative research projects for their valuable contributions towards obtaining the presented research results. In particular, the authors would like to acknowledge the contributions done by the researchers from Technicolor, France, in Chapter 5 of this book.

List of Abbreviations

3D	Three-Dimensional
3DVC	Three-Dimensional Video Coding
3GPP	Third Generation Partnership Project
ACK	Acknowledgement
ACR	Absolute Category Rating
ADSL	Asymmetric Digital Subscriber Line
AVC	Advanced Video Coding
BS	Base Station
CDMA	Code Division Multiple Access
CDN	Content Delivery Networks
CG	Computer Graphics
CGI	Computer Generated Imagery
CTU	Coding Tree Unit
CVR	Comfortable Viewing Range
DAB	Digital Audio Broadcasting
DCR	Degradation Category Rating
DCT	Discrete Cosine Transform
DES	Depth Enhanced Stereo
DMB	Digital Media Broadcasting
DRM	Digital Rights Management
DSCQS	Double Stimulus Continuous Quality Scale
DSL	Digital Subscriber Line
DSLAM	Digital Subscriber Line Access Multiplexer
DTT	Digital Terrestrial Television
DVB	Digital Video Broadcasting
DVB-H	Digital Video Broadcasting-Handheld
DVB-S	Digital Video Broadcasting-Satellite
DVB-T	Digital Video Broadcasting-Terrestrial
EPG	Electronic Program Guide
ETSI	European Telecommunication Standardisations Institute

EU	European Union
FEC	Forward Error Correction
FMO	Flexible Macroblock Ordering
FP7	Seventh Framework Programme of the European Union
FPR	Film-type Patterned Retarder
FOD	Field of Depth
FOV	Field of View
FR	Full Reference
FVV	Free-Viewpoint Video
GPU	Graphical Processing Unit
GSE	Generic Stream Encapsulation
GSM	Global System for Mobile Communications
HD	High Definition
HEVC	High Efficiency Video Coding
HMP	Head Motion Parallax
HSPA	High Speed Packet Access
HVS	Human Visual System
IDR	Instantaneous Decoder Refresh
IEEE	Institute of Electrical and Electronics Engineers
IMDB	Internet Movie DataBase
IMT	International Mobile Telecommunications
IP	Internet Protocol
IPTV	Internet Protocol Television
ISO	International Organization for Standardization
ITU	International Telecommunications Union
JCT	Joint Coding Teams
JND	Just Noticeable Difference
JVT	Joint Video Team
LCD	Liquid Crystal Display
LDI	Layered Depth Image
LDPC	Low Density Parity Check
LDV	Layered Depth Video
LGN	Lateral Geniculate Nucleus
LTE	Long-Term Evolution
MAC	Medium Access Control
MAN	Metropolitan Area Network
MANE	Media Aware Network Element
MB	Macroblock
MBMS	Multimedia Broadcast Multicast Service
MC	Motion Compensation
MDC	Multiple Description Coding
ME	Motion Estimation
MFN	Multiple Frequency Network
MIMO	Multiple Input Multiple Output
MOS	Mean Opinion Score

MPEG	Motion Pictures Experts Group
MPML	Multiple Physical Layer Pipe
MVC	Multi-view Video Coding
MVD	Multi-view Video plus Depth
MVD2	Multi-view Video plus Depth -2 views
MVV	Multi-View Video
NAL	Network Abstraction Layer
NR	No Reference
OFDMA	Orthogonal Frequency Division Multiple Access
P2P	Peer-to-Peer
PSNR	Peak-Signal-to-Noise-Ratio
QAM	Quadrature Amplitude Modulation
QoE	Quality of Experience
QoS	Quality of Service
QPSK	Quadrature Phase Shift Keying
RLC	Radio Link Control
RR	Reduced Reference
RTP	Real-time Transport Protocol
SAO	Sample Adaptive Offset
SD	Standard Definition
SFN	Single Frequency Network
SLM	Spatial Light Modulator
SNR	Signal-to-Noise-Ratio
SSIM	Structural SIMilarity Index
STB	Set-top Box
SVC	Scalable Video Coding
TDD	Time Division Duplex
UDP	User Datagram Protocol
VCEG	Video Coding Experts Group
VDSL	Very high bit-rate Digital Subscriber Line
VLC	Variable Length Coding
VoD	Video on Demand
VQEG	Video Quality Experts Group
VQM	Video Quality Metric
WiMAX	Worldwide Interoperability for Microwave Access

1

Introduction

Recent years have seen a reawakening of interest in 3-dimensional (3D) visual technology. 3D, in the form of stereoscopy, has been with us since 1838, when it was first described by Sir Charles Wheatstone. Since then there have been a number of periods when interest in 3D technology has surged and then faded away again. Each resurgence in interest can largely be put down to the development of new technologies, or new marketing initiatives. The constant reawakening of interest also demonstrates the strong desire of the public for immersive 3D experiences. The fading away of interest can largely be put down to the disappointing nature of previous generations of 3D technology.

We are currently at the beginning of another resurgence in interest in 3D, which is likely to be durable. There are a number of reasons why this should be:

- affordable, aesthetically pleasing, 3D displays, which are as capable of displaying high quality 2D colour video, as they are of showing high quality 3D video;
- digital video production techniques to allow correction and optimization of captured 3D video during post-production;
- new developments in the understanding of 3D perception, which enable the production of content which is more comfortable to the eye;
- new formats and standards for the compression of 3D video in digital formats, enabling high quality 3DTV services to become a reality.

This book aims to provide the reader with an overview of the key technologies behind the current generation of 3D technologies, and also to provide a guide to where the technology will head next. It covers the full chain, from capture of 3D video, to display. In between, it examines issues such as 3D video

3DTV: Processing and Transmission of 3D Video Signals, First Edition.
Anil Fernando, Stewart T. Worrall and Erhan Ekmekcioğlu.

compression, assessment of 3D video quality, and transmission of the video over a variety of networks.

In this chapter, Section 1.1 describes the history of 3D video, highlighting the key developments since the nineteenth century. Section 1.2 describes the most common digital 3D video formats currently in use. The motivation for the book, and for the reintroduction of 3D video in general, is outlined in Section 1.4. The most common application scenarios for 3D video are discussed in Section 1.3. Finally, Section 1.5 gives an overview of each of the book chapters.

1.1 History of 3D Video

Before examining the current state-of-the-art in 3D visual related technology, it is instructive to examine the way in which 3D technology has developed, and the reasons for previous failures. In this way it is possible to assess the durability of the current 3D boom, and to consider which of the remaining challenges are the most important to solve.

Figure 1.1 gives a summary of some of the most important milestones over the past 150 years. One aspect that may surprise some readers is the length of the timeline. Many key developments took place either in the nineteenth century or the early twentieth century. 3D was most popular during the 1950s and 1980s, but each 3D boom faded within a few years. The following subsections describe the key technological developments and discuss the reasons for the promotion and subsequent failure of 3D movie technology. Finally, the latest resurgence in popularity in 3D is examined. It is important to note that it is only relatively recently that 3D displays for the home have been available at an affordable price.

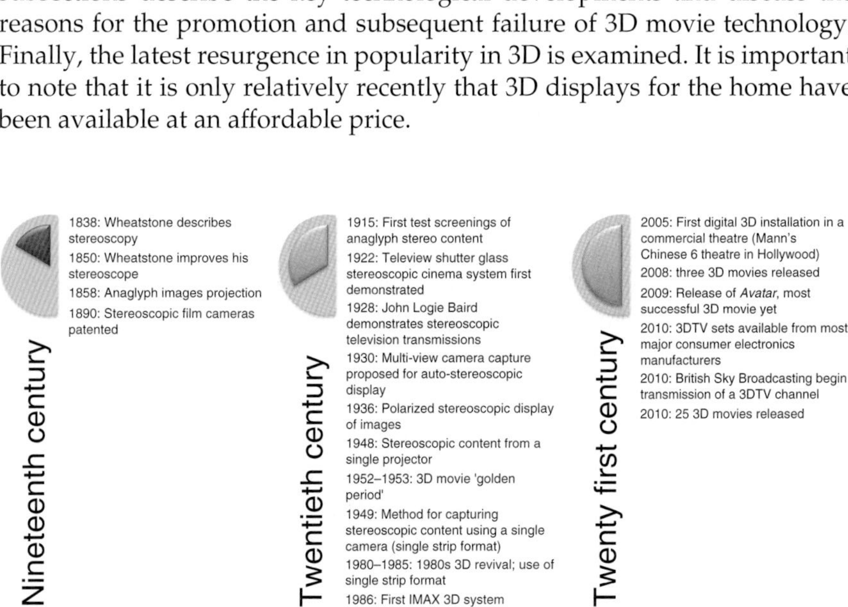

Figure 1.1 Key developments in the history of 3D video

This section of the book describes the history of 3D video-related technology and applications over time, up until the twenty-first century. There follows specific sections on auto-stereoscopic displays, and 3DTV-specific developments. Auto-stereoscopic and volumetric displays have been treated separately because they have been of considerable interest to the research community, but have not yet been commercially exploited.

The history described here is relatively brief, and some details have been left out. For example, a number of significant developments were made in the Soviet Union, which are not covered here. For a more detailed historical overview, readers should consult other references [1, 2].

1.1.1 3D in the Nineteenth Century

Sir Charles Wheatstone is widely considered to be the father of stereoscopy. In 1838, he published a paper describing how each eye sees a slightly different version of the same scene [3]. It seems unlikely that Wheatstone was the first person in history to notice this effect. Wheatstone himself provides a quote in his 1838 paper, which suggests that Leonardo Da Vinci would probably have been aware of the effects of binocular vision. Other authors have noted that Euclid also made certain observations about binocular vision [1]. However, Wheatstone is the first to describe stereoscopy explicitly and in detail. Wheatstone's 1838 paper also described the stereoscope, which allowed viewing of stereoscopic drawings. As shown in Figure 1.2, The stereoscope used mirrors to project the drawings at E′ and E to the position of the viewers' eyes (A′ and A respectively). In 1840, he was awarded the Royal Medal of the Royal Society for his work on stereoscopy.

The next notable development was the introduction of anaglyph images. The anaglyph approach involves the use of glasses where the two lenses are different colours. Red-cyan lenses were often used for viewing stereoscopic

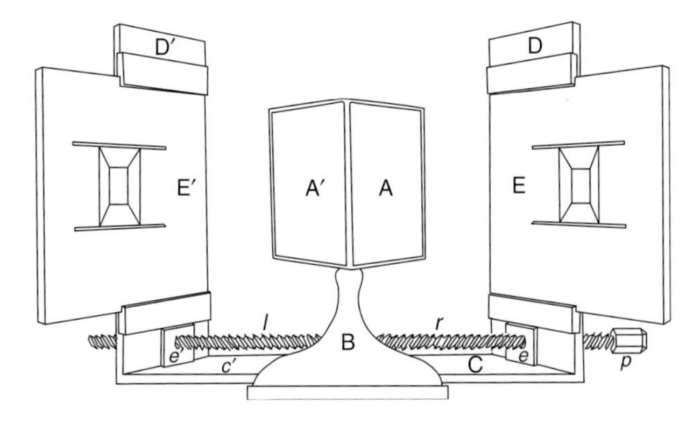

Figure 1.2 Wheatstone's mirror stereoscope, taken from his 1838 paper, which allows viewing of a stereo pair of drawings (3) (Reproduced with permission of the Royal Society Publishing)

images. The two stereoscopic views are superimposed, and the colour lenses are used to filter out one of the views so that the left eye only sees the left view, and the right eye only sees the right view. There are three key milestones during the development of anaglyph image technology. The first milestone is the publication of Wilhelm Rollmann's paper, which described the anaglyph principle from experiments involving red and blue lines, viewed using red and blue lenses [4]. In 1858, Joseph D'Almeida developed a system to project two images, through red and blue lenses, onto a single screen [1]. The images could then be viewed using glasses with red and blue lenses. The third key milestone was reached by Louis Du Hauron, who was at the leading edge of the development of colour photography. Du Hauron proposed a method for combining the two stereoscopic views on to a single print [1]. This eliminated the need for two lenses in projectors, and also enabled printed anaglyph images to be produced.

Stereoscopic images and drawings became quite popular during the nineteenth century. However, the development of technology allowing the capture and display of 3D movies proved to be highly challenging. Of course, much of this can be put down to the primitive nature of motion picture cameras (2D or 3D). Researchers and inventors were still struggling to produce reliable cameras capable of filming at rates greater than a few frames per second. However, some efforts were made, most notably by Frederick Varley and William Friese-Greene. Friese-Greene is a particularly notable figure, given the quotation on his tomb:

> His genius bestowed upon humanity the boon of commercial cinematography of which he was the first inventor and patentee.

Research by Brian Coe has subsequently shown that this claim is something of an exaggeration [5]. Friese-Greene was certainly a very active inventor, filing large numbers of patents. However, many of his inventions proved to be either impractical or unreliable. This description at least partially matches his stereoscopic camera, which he developed in collaboration with Frederick Varley in 1890. By all accounts, the camera was unreliable and only capable of capturing at a rate of a few frames per second, which is not enough to create a true sensation of movement. Furthermore, there is no record of any of the captured content being projected or displayed in a practical manner. Shortly after this development, Friese-Greene was declared bankrupt, and the realization of a commercially viable 3D system was not achieved until the next century.

1.1.2 Early Twentieth-Century Developments

The early twentieth century saw the arrival of a number of key technologies, the basic principles of which are still used in many of today's 3D technologies.

Shutter glasses, and polarized stereoscopic viewing technologies were all initially developed during this period. Shutter glasses and polarized lenses of course form the basis of the stereoscopic systems in use today. However, many initial developments were made using anaglyph systems.

One of the most reliable first appearances of 3D technology in the twentieth century is described by Lynde in a 1915 article in *The Moving Picture World* [6]. Lynde reports a demonstration of a new red–green anaglyphic stereoscopic movie projection system. This example is cited not just because of the reliability of the source, but also because it involved a major Hollywood director. Edwin S. Porter was a well-known movie maker, having directed one of the most important and popular silent movies, namely *The Great Train Robbery*. His presence therefore added a great deal of credibility to the event. The demonstration was made by Porter and the co-inventor of the new system, William E. Waddell, on June 10, 1915, in the Astor Theater, New York City. During the 1915 demonstration, a number of short features were shown, which were filmed as images 2½ inches apart. Two projectors were used to display the left and right views in the red–green anaglyph format. Reaction to the demonstration appears to have been mixed, with Lynde claiming:

> Images shimmered like reflections on a lake and in its present form the method couldn't be commercial because it detracts from the plot.

Elsewhere in his article, his descriptions suggest that there were significant problems with synchronization between the two projectors, resulting in significant eye strain among the audience.

The next significant development was presented in December 1922 at the Selwyn Theatre in New York City. Here a new system called *Teleview* was presented by the inventors Laurens Hammond and William F. Cassidy. Their invention made use of what would now be called 'shutter glasses' or 'active stereo' technology. Two reels of film, representing the left and right eye views, were run through two separate, but synchronized projectors. One reel, however, was intended to be one frame behind the other, meaning that the left and right eye views were projected alternately onto the screen. Audience members sat behind their own viewing device, which featured a mechanical shutter, driven by a motor. The motor was synchronized to the projectors. The shutter blocked the left eye when the right eye view was displayed on the screen and vice versa. The 3D effect was reportedly much better than the 1915 demonstration described by Lynde [7]. An article in *The New York Times* stated:

> ... those who went to the Selwyn last night were surprised, sometimes startled and often delighted with the vividness of the pictures that they saw and the unusual effects obtained by the use of the Teleview device.

However, the high cost of the equipment and the lack of attractive movie content meant that the system was not widely adopted.

Anaglyph filters tend to distort the colours perceived by the viewer, and are therefore not a good way of viewing high quality 3D images of video. The introduction of polarizing filters therefore provided a huge leap forward in stereoscopic video quality. Polarization sheets were first produced by William Bird Herapath, who discovered a method of forming polarizing crystals in 1852 [8]. However, the sheets, formed from Herapathite, proved to be of low quality, which limited their practical use for viewing stereoscopic images. Nevertheless, according to Lipton [1], John Anderton proposed the use of polarizing sheets for stereoscopic display in a patent, filed in 1891. The 1891 British patent is difficult to find. However, a US patent filed in 1893 refers to the original British version [9]. It is also possible to find Anderton's 1898 British patent for stereoscopy using polarizing filters [10]. This 1898 patent states that it provides improvements on the earlier 1891 British patent.

Edwin H. Land improved upon Herapath's original work, and filed the first of his many patents in 1929 [11]. Land was the man who founded the Polaroid company, and was also responsible for the invention of the Polaroid instant camera. In January 1936, he gave the first demonstration of 3D projection using polarizing filters at the Waldorf-Astoria Hotel [12]. During projection, two reels, carrying the right and left eye views respectively, had to be synchronized using an external motor. Another complication was that polarized light would not register on a normal matte white screen, and a silver screen was required to correctly display the two views. Despite these complexities, the quality appears to have been favourable according to the *New York Times* article:

> Observers were ushered into a seemingly living fairyland of forms and colours.

Land followed the first demonstration with another at the New York Museum of Science. It is this kind of polarizing technology that formed the basis for the systems that were used during the first 3D boom in the 1950s.

1.1.3 The 1950s 'Golden' Period

Although there was considerable interest by scientists and researchers in developing 3D movie technology, it was not until the early 1950s that the movie studios started to take a serious interest in the process. Movie studio executives were becoming increasingly concerned about the impact of television on cinema audiences. Between 1946 and 1952, weekly attendance figures in US cinemas had fallen from 82.4 million to 46 million [1]. In addition to this, the House Un-American Activities Committee (HUAC) had blacklisted talented scriptwriters, and directors [13]. The US movie industry was in crisis and needed to find something to attract people back into the cinemas. This dire situation led to experimentation with new technology, so that cinemas could provide an experience far superior to television.

According to Lipton [1], the first American movie to be made in colour and 3D was *Bwana Devil*, which was first screened on November 27, 1952. The movie was a hit, and grossed $100,000 in its first week. This woke the industry up to the potential of 3D, and led to the production of more 3D movies by the major studios. The production process was significantly limited by the fact that most of the major studios had neither the 3D cameras rigs, nor the expertise to shoot movies in 3D. This meant that no more than 45 3D movies were produced in a single year, compared to the typical 2D output of around three hundred per year. It is interesting to note that the entire 3D movie boom only lasted around nine months. By this time, the public had shown a clear preference for seeing 2D movies.

A number of reasons have been put forward for the failure of the 1950s 3D boom. One theory is that the quality of the feature films was particularly poor. Although 3D poor movies were made, there were a number of critically acclaimed movies made, such as *Dial M for Murder*, *House of Wax*, and *Kiss Me Kate*. Table 1.1 shows a selection of 3D movies from the 1950s, along with the average ratings given by users of the Internet Movie Database (IMDB). This reveals that while there were some very poor films made, there are also examples of excellence. It cannot be said that the quality of 3D movies was, on average, any worse than 2D features during the same period.

The reason for the rapid disillusionment by the public is most likely to be the significant eyestrain that many suffered after relatively short periods in the cinema. The eyestrain itself can be put down to a number of factors:

- *A lack of stereoscopic shooting expertise among film crews and inadequacies in understanding of optical problems* – Film crews in the 1950s had to learn about shooting in stereo very quickly. One of the key issues in avoiding eyestrain is to choose the optimal interaxial distance for the lenses. Too great a distance will result in eyestrain. The human interocular distance is typically 2½ inches, so distances greater than this are likely to be too much for viewers. Some rigs had a minimum interaxial distance of 3½ inches, which would have caused problems for some cinemagoers. In addition to this, there were sometimes shifts in the position of the optical axes of the lenses. In the worst cases, this led to vertical parallaxes. Even small vertical parallaxes cause significant muscular strain, as one eye is forced to look slightly upwards compared to the other in order to achieve fusion of the two views.
- *Poor quality control during film processing* – Another factor that can cause eyestrain over time is differences between the left and right views. If the two views are not processed in exactly the same way, then one view may be lighter than the other.
- *Projection systems too complex for most projectionists to handle* – During the 1950s, projectionists were not always able to ensure that the equipment was

Table 1.1 Selection of film titles released during the 1950s 3D boom, with movie ratings taken from IMDB

Title	Year	IMDB Rating
The French Line	1953	5.1/10
Taza, Son of Cochise	1954	5.5/10
Creature from the Black Lagoon	1954	6.9/10
Dial M for Murder	1954	8.1/10
Bwana Devil	1952	5.2/10
House of Wax	1953	7.0/10
Man in the Dark	1953	6.2/10
It Came from Outer Space	1953	6.6/10
Kiss Me Kate	1953	7.2/10
Hondo	1953	7.1/10
Miss Sadie Thompson	1953	6.0/10

correctly set up. Typical problems included differences in projection lens focal length, and differences in illumination of the two views. Furthermore, if films became damaged, it was common practice for projectionists to remove damaged frames. If the damaged frames were not removed from both views, then temporal synchronization would be lost. Also, if we consider that approximately 8% of the population cannot perceive stereo, then there is a risk that the projectionist might not have had stereo vision, making it very difficult for them to configure everything successfully. According to Lipton [1], Polaroid conducted a survey of one hundred stereo-equipped cinemas during 1953. The survey found that 25 of those theatres were poorly set up, causing significant eyestrain among viewers.

- *Corner cutting by cinemas* – The cinemas attempted to reduce their costs by using inferior screen coatings, and by using cheap polarizing glasses. This meant that images were not as bright and sharp as they should have been, and that the viewing experience using the glasses was poor.

When all of these difficulties became apparent, attention switched to alternative technologies for persuading television viewers to come to the cinema. The principal enhancement that the studios considered was changing the aspect ratio, making the screen wider. Examples of this technology include CinemaScope, which changed the standard 1.37:1 aspect ratio to 2.66:1.

1.1.4 The 1980s Revival and the Arrival of IMAX

Although the first 3D boom proved to be disappointingly short-lived, this was not the end of interest in 3D from the public or the movie industry. Certainly by the late 1970s, the movie industry was again concerned about new technology making home viewing more attractive than visiting a cinema. Mass market Video Cassette Recorders (VCR) had arrived, which allowed the public to watch movies at home repeatedly, and at a time of their choosing. This was perceived as a clear threat to movie industry revenues. However, it was clear from the problems described in Section 1.1.3 that more development work was required to make 3D technology commercially viable. Researchers continued to improve the technology, focusing on techniques that would provide a solution to problems such as view synchronization.

The Polaroid Vectograph was one of the candidates offering to provide a solution to view synchronization. It was invented by Edwin Land and Joseph Mahler, and allowed the two views to be placed on one film strip, rather than two [14]. The film was double-sided, and acted as a polarizing filter, as shown in Figure 1.3. This meant that no polarising filters were required for the projectors, resulting in a brighter image. The fact that the whole frame was used for each image also meant that there was no loss of resolution, and combining the two views on one film strip meant that there were no problems in synchronizing the two views. The original Vectograph worked on still images, and further development was needed to achieve motion picture capability. Shortly after the still image Vectograph was patented, Land patented a technique for a Vectograph capable of shooting movies [15]. However, by the time the technology was ready for commercial use, the 1950s boom was over, leaving Land with a promising product, but a non-existent market.

Colonel Robert Bernier developed techniques for projecting stereoscopic movies from a single projector [16], which is illustrated in Figure 1.4. Later, he developed a method for filming with a single camera, where the two views are projected onto the same frame of a standard camera film [17]. Like the Vectograph, Bernier's system solved the problem of view synchronization loss. As both views were shot onto a single strip of film, problems with film processing could also be avoided. Bernier's system was called SpaceVision, and was used to shoot a number of movies in 3D. In fact, it was systems similar to SpaceVision that were used during the 1980s 3D revival. It is interesting to note a number of disadvantages of systems such as SpaceVision:

- *Loss of image resolution* – The SpaceVision system requires that two views are projected onto a single frame of standard film. Inevitably, the physically smaller area per frame led to a loss in picture fidelity.
- *Loss of image brightness* – Unlike the Vectograph system, SpaceVision requires one set of polarizing lenses in front of the projectors, and another set to the used in the viewing glasses. Two sets of polarizing lenses led to the loss of a significant amount of light.

A number of alternative systems to SpaceVision were developed. For example, the Stereovision system placed the two views side-by-side on a single strip of film. Unlike SpaceVision, a 65mm print size was used,

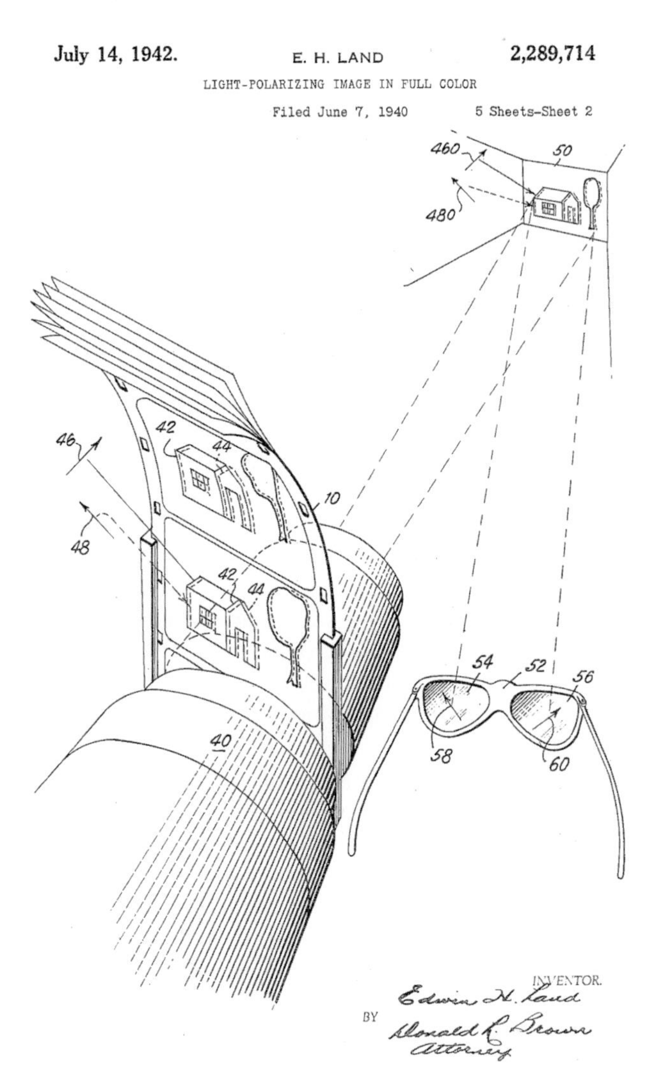

Figure 1.3 Diagram illustrating the basic principle of projection of Vectograph film, taken from Land's 1940 US Patent (14) (Reproduced with permission of the Polaroid Corporation)

meaning higher quality prints. However, for projection, it was often still necessary to reduce print to 35mm, meaning that the quality was not much different from that obtained from SpaceVision.

Although the 3D experience of the 1980s was an improvement over the 1950s' one, the inherent limitations of the SpaceVision type systems meant that 3D pictures were more blurry and were darker than for the same 2D movie. This limited the attraction of 3D for the general public. In addition to these inherent problems with the SpaceVision approach, it is clear from

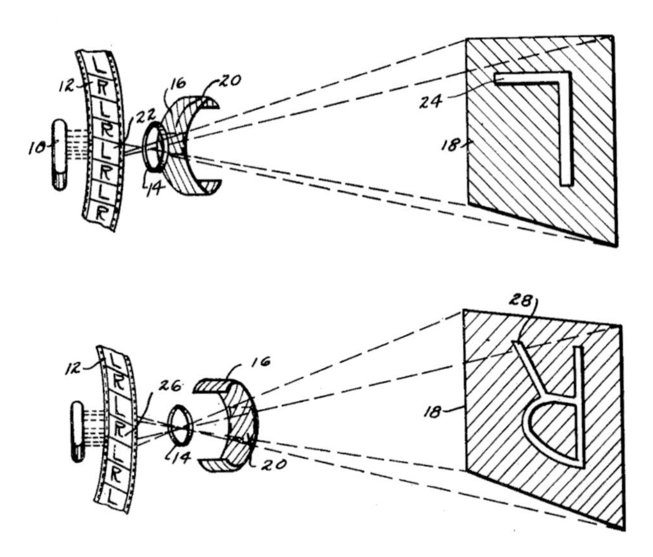

Figure 1.4 Illustration of Bernier's polarized projection system, taken from his 1949 US Patent (16). Left and right frames are contained on the same strip. A polarizing filter is flipped in synchronization with the frames as they pass through the projector (Source: US patent 2478891 (http://www.google.com /patents/US2478891))

expert observers' opinion that some films were poorly shot and processed. This quote from Lipton [1] about one production of the era is damning:

> The summer of 1981 saw the Filmways release of a western shot in Spain, titled *Comin' at Ya*. Production values were low, the acting was terrible, the dialogue moronic, the stereoscopic process, Optrix, was an optical catastrophe, and the filmmakers attempted to place every shot behind the heads of the audience. The stereoscopic system suffers from left and right images of unequal sharpness, severe vertical parallax, and strange watermark-like spots hovering in one or the other image field. The film has apparently been doing good business, and for that reason many major studios and producers are considering employing the three-dimensional medium. The situation brings to mind *Bwana Devil*.

Although not all of the productions were as bad as described by Lipton, in general, the quality problems meant that the new 3D revival was once again short-lived. Only a limited number of 3D movies were shot during the early 1980s, and the technology once again fell out of favour. One conclusion that could be drawn from this revival, is that for 3D to be a successful and enduring fixture, it needs to provide an experience that is not worse than that of 2D.

Development of 3D technology continued. One notable development was the Stereospace system, which was put together by Richard Vetter for United Artists Communications. The system provided a step forward in terms of quality by using two 65mm reels, with the left and right views on separate reels [18]. A magnetic strip with timing information was used to maintain synchronization between the two projectors required for presentation. A similar system to Stereospace was developed by Disney for inclusion in their theme parks. Disney's system was more than a straightforward stereo 3D projection system though. In addition to 3D, in-theatre effects were included, such as laser shows and smoke. One of the most notable 3D Disney productions was *Captain EO*, which starred Michael Jackson, and was directed by Francis Ford Coppola. Both the Stereospace and Disney systems were prone to jitter and vertical parallax. However, Disney's system proved to be more successful and enduring. This was in part because Disney controlled everything from film production to projection, ensuring that quality control was tightly enforced.

The system that became most widespread in the 1980s was IMAX. Special IMAX cinemas have now been set up all around the world by the IMAX Corporation. The company control all aspects of the process, from producing cameras to capture on to 70mm film, to the projection system in each theatre. Early IMAX systems were installed in the 1970s, while the first 3D systems were rolled out in 1986. The IMAX system provides high quality 2D colour picture quality on very large, immersive screens. Filming for IMAX 3D is carried out using special cameras, where lenses 2½ inches apart are used to capture material onto two different films. Two synchronized projectors are used, which have polarizing filters placed over their lenses. The movie is then viewed using glasses with polarized lenses. The only issue with IMAX theatres is that they are not currently as widely available as traditional cinemas. It has managed to carve out its own niche, but movies are still seen by most people in standard format cinema facilities.

1.1.5 The Twenty-first-Century Revival

The twenty-first century has seen a significant revival in 3D technology. Once again, the movie industry has taken the lead by reintroducing 3D into cinemas. Although – thanks to IMAX – 3D technology has never really gone away, the twenty-first century has seen more widespread adoption of the technology by mainstream cinemas. New threats are being faced by the movie industry, in the form of piracy. File sharing technologies have enabled new movies to be shared, and freely downloaded. This can sometimes occur before the official release date. In addition, whereas pirate copies used to be low quality, modern pirate movies can often be obtained in HD. Therefore, providing high quality 3D cinema releases is a way of giving consumers something that cannot be experienced at home. It also acts as an effective method of preventing illegal recordings taking place by bootleggers who take hidden cameras into the cinema.

Technology has improved significantly since the first 1950s boom, although there are still significant gaps in knowledge when it comes to 3D quality (see Chapter 6 for a discussion of 3D quality issues). The new generation of 3D movies have experienced significant commercial success. Currently, three of the top 10 grossing movies of all time have been released in 3D.[1] Of course, these films might have been as successful in 2D as they are in 3D. However, some market research by the International 3D Society has suggested that 3D movies have taken more profits in their 3D form, despite there being fewer 3D screens available [19]. Care should be taken with these figures, given that 3D movie tickets are more expensive than 2D movie tickets. However, the feedback obtained from consumers in the same research is very encouraging, with 74% of 3D movie viewers saying that 3D movies are better than 2D.

1.1.6 Auto-Stereoscopic

One of the inherent problems with many 3D display technologies is that they require the viewer to wear special glasses. Some viewers find such glasses off-putting, and therefore researchers have put considerable efforts into developing display technologies that do not require glasses. Such displays usually fall into the auto-stereoscopic category, which is described in this subsection.

Two main classes of auto-stereoscopic have been deployed:

- *Parallax barrier* – where a barrier is placed in front of the display. The barrier features a series of regularly spaced slits, which ensure that each eye sees a different area of the screen. The left and right views are carefully spliced together before display, so that when viewed, a 3D binocular effect is obtained. The problem with this technology is that it is extremely sensitive to head movements. Even a small head movement can lead to unpleasant effects, such as reverse stereo where the left eye sees the right eye view and vice versa.
- *Lenticular lens* – where the display is coated with an array of semi-cylindrical lenses. An image strip sits behind each lenticule, which contains a succession of partial views of the subject, from the extreme left camera position to the extreme right. Lenticular lenses are more expensive to produce than parallax barriers, but allow greater head movement by the viewer before unpleasant stereo artefacts can be seen.

These display types are discussed in more detail in Section 5.4.4. Here, we are concerned mainly with the historical progress of these technologies.

[1] As published by http://www.boxofficemojo.com on 21 February 2011.

One of the earliest publications describing the parallax barrier concept was produced by Auguste Berthier in 1896 [20]. Further developments were reported by Frederick Ives in 1902 [21, 22]. Clearly, at this stage, many of the developments were made using drawings, or still images. Motion picture cameras during this period were still relatively limited in their capability.

The credit for proposing lenticular lenses is often given to Gabriel Lippmann for his paper published in 1908 [23]. Once again, the technology was mainly limited to use on still images at this point. In 1930, Herbert Ives filed his patent describing a system for auto-stereoscopic projection of motion pictures, based on lenticular lenses [24]. The system principles are illustrated in Figure 1.5. As can be seen from Figure 1.5, the arrangement made use of multiple cameras and multiple projectors, which all required precise synchronization. It is difficult to find evidence of a successful implementation of the ideas put forward in this patent, and it seems likely that this was an idea that remained on the drawing board.

Over the years, many proposals for auto-stereoscopic displays have been put forward [25–29]. The result has been a steady and continual improvement in the quality of the displayed video, with increases in resolution, and the number of available viewpoints. Of course, the technology has not been fully exploited for cinema, because of the restrictions of seating position and head movement required for viewing stereoscopic video. There are now a large number of companies offering auto-stereoscopic displays, including Alioscopy, Dimension Technologies, and NewSight. The quality of these displays is very high, but they cost more than similar-sized displays that make use of shutter glasses. This and the head movement restrictions have so far prevented significant commercial success for this type of technology.

1.1.7 3D Television Broadcasts

Many of the historical developments described in this section have been targeted at providing 3D cinema experiences. Of course, there is a significant amount of overlap between cinema and television. However, television pictures need to be transmitted, and then shown on displays that are affordable for the average consumer. This section therefore describes some of the developments towards providing actual 3D television services for consumers.

One of the earliest examples of a 3D television was put together and demonstrated by John Logie Baird. He filed his first stereoscopic television patent in 1926 [30], and demonstrated the system in 1928 [31]. Transmission of two views of a person's face were demonstrated within the laboratory. The two views were alternately shown on the left and right sides of a single neon tube. A stereoscope device was used by viewers, which deployed a prism to direct the views to the eyes and therefore allow fusion. Of course, this system was a long way from being a practical solution. There were still many quality issues to solve with non-stereoscopic television, and

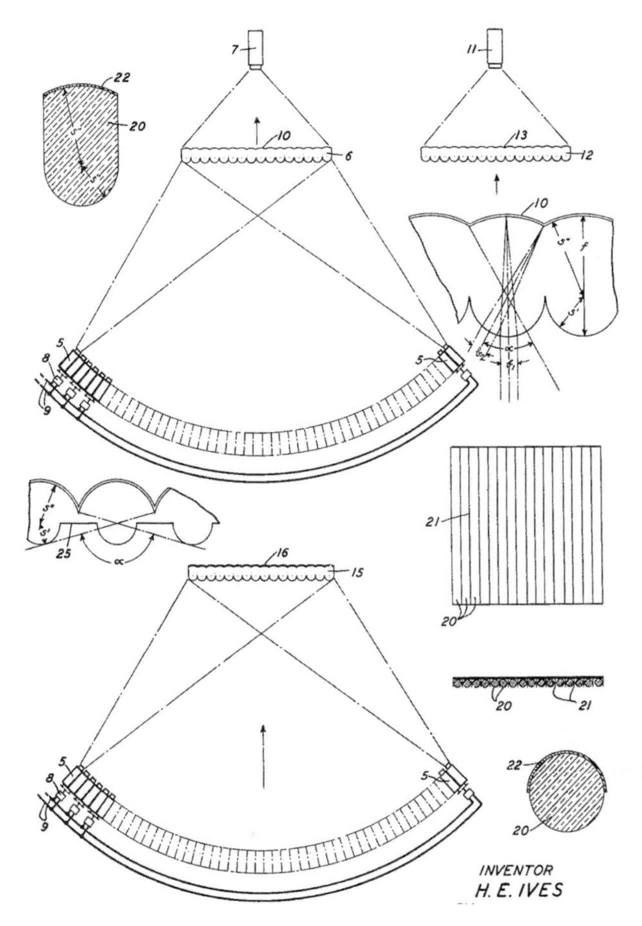

Figure 1.5 Drawings taken from Ives' patent for auto-stereoscopic projection (24). The drawings assume that fifty cameras were originally used during filming. In Fig. 1, one projector per original camera is used to project the views onto a translucent lenticular array. Light passes through this array to be captured by a single camera. In Fig. 2, material captured by the single camera is projected onto another lenticular array for viewing. Fig. 3 shows an alternative viewing scenario, where the fifty projectors are used in conjunction with a lenticular array backed with a white screen (Reprinted with the permission of Alcatel-Lucent USA Inc.)

the stereoscope viewing method was only suitable for a single viewer at a time.

The next 3DTV event of note was the experimental broadcast in the US on April 29, 1953. A trial broadcast of the series *Space Patrol* was conducted in Los Angeles at the National Association of Radio and Television Broadcasters 31st Annual Gathering. The experimental broadcast used polarizing technology to display the stereoscopic content [2]. The trial was conducted at the same time as the 1950s boom, to which television broadcasters were almost certainly paying attention. The problem with the technology based on the polarizing glasses was that it required consumers to purchase special expensive new television sets. This, and the rapid demise of 3D cinema in the 1950s, meant that interest in 3DTV died out.

Further broadcasts were made in the 1980s, following the introduction of a new system developed by Daniel Symmes, founder of the 3D Video Corporation. This new system allowed anaglyph format stereoscopic video to be broadcast and displayed on standard television systems. The system is described in patents filed in various countries, including Britain [32]. This particular patent includes diagrams demonstrating how it can be used with a National Television System Committee (NTSC) television system. Use of standard television sets is the key advantage of this approach. Consumers only need to obtain anaglyph glasses, which can be produced relatively cheaply. Although broadcasts in the US and Europe were initially popular, interest died out as consumers realized the limitations of anaglyph approaches (i.e. poor colour rendition).

During the 1980s, the Institut für Rundfunktechnik (IRT) worked on broadcast systems that relied on polarizing glasses. Systems were demonstrated at a number of trade fairs [33], and received very good responses from those who saw the technology. At the trade fairs, two projectors were used for the two polarized stereoscopic views. In addition to this system, IRT produced a smaller polarized display using standard television monitors, mirrors, and polarizing filters. Sand claimed that one of the limitations was a lack of satisfactory 3DTV cameras [33]. The viewing equipment would also have been significantly more expensive for consumers than a standard television set.

Some attempts were made in the 1980s and early 1990s to exploit the Pulfrich effect. The Pulfrich effect involves the use of special glasses, where one lens is darker than the other [34]. This reduced illumination seems to cause a delay in signal transmission to the brain, and therefore the difference between the two eyes results in a perceived depth disparity. The effects are fairly limited, as 3D can only be perceived for objects moving left of right across the screen. The direction of their depth change depends on the direction of movement of the object *and* the eye with the darkened lens. 3D effects are not seen if movement is vertical, with respect to the camera, or if the object moves towards or away from the camera. This effect has limited value for true 3D reproductions, and its use has largely been restricted to short sections of television programmes or advertising. The restriction on movement is too great for practical widespread use in 3DTV production.

Digital television standards were introduced during the 1990s, which largely relied upon the MPEG-2 video codec for compression. Although this codec could not originally handle stereoscopic video, it was updated to include the Multi-View Profile (MVP) [35]. However, this profile has not seen commercial usage so far, and is unlikely to do so. More recent standards are capable of coding multiple views with much greater efficiency than this MPEG-2 extension (see Section 3.2 for more details).

Significant 3DTV work has been carried out in Japan over the years. Sand's (1992) paper describes research at the University of Tokyo, where an auto-stereoscopic system with up to 24 channel multiple-views was displayed [33]. Nippon Hōsō Kyōkai (NHK) developed a 3D HDTV relay system, which was used in 1998 to transmit stereoscopic live images of the Nagano Winter Games via satellite to a large-scale demonstration venue in Tokyo [36].

The recent arrival of affordable 3D television sets means that 3DTV broadcasts are once again of interest to broadcasters. Many experimental broadcasts are taking place across the world, particularly using the frame compatible stereoscopic format (see Section 1.2.1). The frame compatible format can already be deployed using existing digital broadcast technology, such as DVB-S2. A commercial 3DTV channel was started by Sky Television in the United Kingdom in October 2010.

1.2 3D Video Formats

This section provides an overview of the most common video formats currently in use, or currently being considered by the research community. The formats are summarized in Table 1.2, which also describes the respective advantages and disadvantages. The frame compatible and service compatible formats are currently the most commonly used for commercial applications, and are supported by most consumer 3DTVs. The stereoscopic video formats are relatively straightforward, as they consist of only two colour video views. Therefore, most space in this section is given over to the description of the depth-based formats, and multi-view formats.

1.2.1 Frame Compatible and Service Compatible Stereoscopic Video

The first formats to be used in modern 3DTV systems will be stereoscopic, as this type of video does not require excessive transmission bit-rates, and is easier to capture than the other formats described in this section. All stereoscopic formats suffer from a lack of flexibility in terms of rendering. It is very difficult to change the users viewpoint, and to change the disparity between the views presented to the user. This prevents users from adjusting the presentation of the 3D video for comfortable viewing.

The *frame compatible* stereoscopic video format has been selected for use in the first generation of 3DTV systems. Its principal advantage is that the two views are packed together within a single video frame. This means that the format is compatible with most existing digital television systems, as no change of resolution or frame rate is required to support 3D.

Table 1.2 Summary of 3D video formats and their respective advantages and disadvantages

Format name	Description	Advantages	Disadvantages
Frame compatible stereoscopic video	Two views are combined into a single view by down-sampling horizontally and placing side-by-side	Compatible with existing 2D transmission systems, video decoders and 3DTVs	Involves loss of spatial resolution
Service compatible stereoscopic video	Two stereo views are transmitted jointly, but in separate frames, e.g. in consecutive frames	Compatible with existing 2D transmission systems, video decoders and 3DTVs. No spatial resolution loss	Double the bit-rate of frame compatible stereo
Colour-plus-depth	The depth is used to render two stereo views for the display	Depth can be compressed to a fraction of the size of a colour view	A single colour view means that occluded areas are unavailable during rendering
Multi-view video	Multiple views are captured usually from multiple cameras arranged in an array	Provides support for multi-view and holographic displays, in addition to free-viewpoint applications	Large bit-rate requirements. Depth information must be estimated at the renderer
Multi-view video plus depth	Similar to multi-view video, but with added depth information	Depth information makes rendering easier, and can mean fewer colour views are needed	Accurate depth information may be hard to generate
Layered depth video	Similar to colour plus depth, but with additional colour and depth data to compensate for occlusions	Improved quality over colour plus depth video	Additional band-width required

The disadvantage is that some loss of resolution is implicit, as the original views must be downsampled so that they may be packed into a single frame. Figure 1.6 shows some of the most common frame compatible formats. The side-by-side format was used in the first 3D broadcasts by Sky in the United Kingdom, and most 3DTVs support the format (i.e. they are able to accept side-by-side video as an input and render it as stereoscopic 3D video).

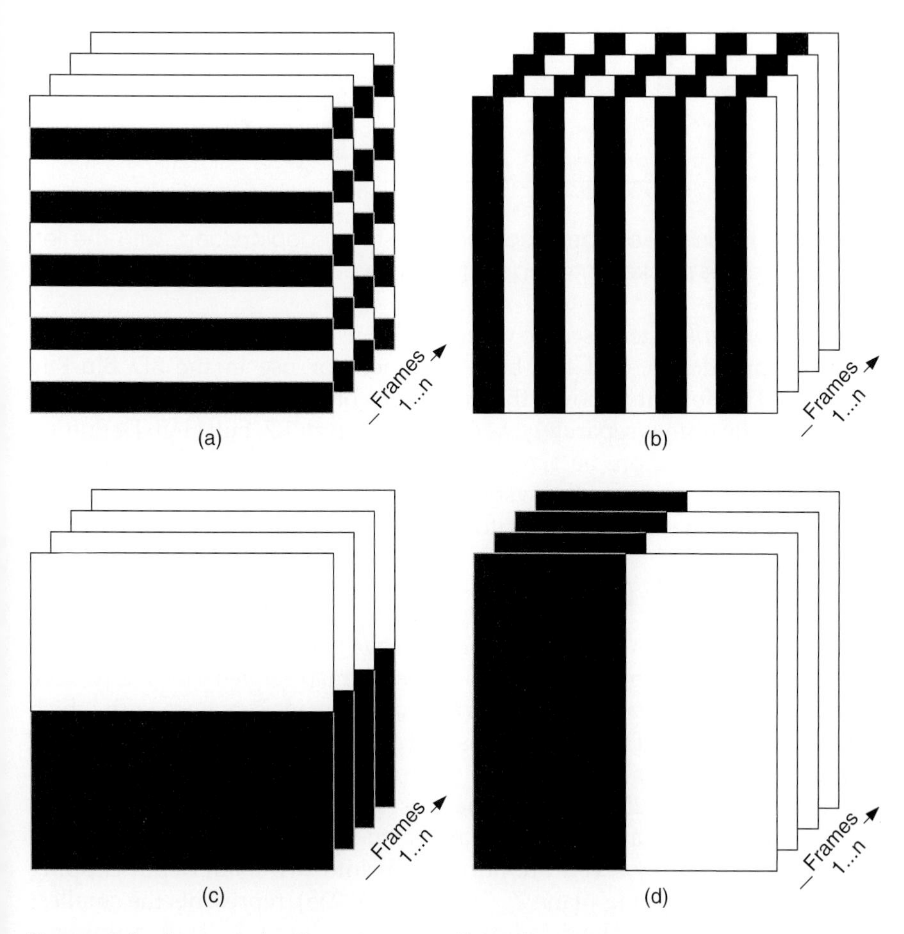

Figure 1.6 Formats for frame compatible stereoscopic video, where black represents pixels from the left view, and white represents pixels from the right view. All arrangements require downsampling of the original left and right views for packing within one frame. (a) Left and right view pixel lines are spliced together. (b) Left and right view pixel columns are spliced together. (c) Left and right views are placed at the top and bottom of the frame. (d) Left and right views are placed side-by-side within the frame

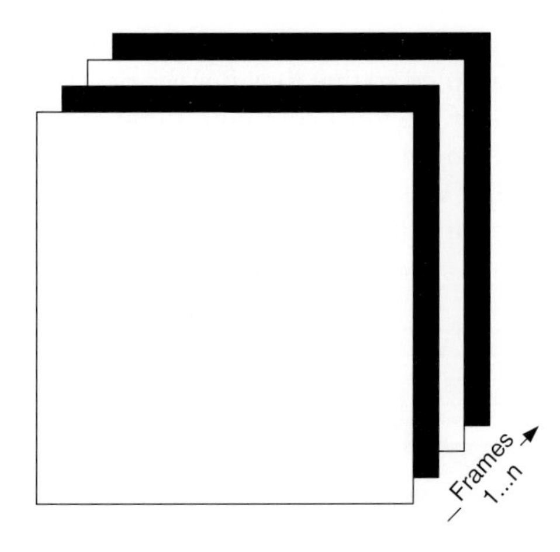

Figure 1.7 Example of frame compatible stereoscopic video, with the left and right frames packed into sequential frames

Service compatible stereoscopic video will be used in the second generation of 3DTV broadcasts, and has been specified for use in the 3D Blu-Ray standard. This format removes the limitations of frame compatible video by presenting the frames separately, as shown in Figure 1.7. Full High Definition resolution can therefore be achieved with this format. Its disadvantage is that it requires changes to the system used to transport the video, requiring users to be supplied with new set-top boxes or disc players.

1.2.2 Colour-Plus-Depth

Depth maps (also known as range images) have been of considerable interest to 3D video researchers in recent years. Depth map sequences usually have similar spatio-temporal resolution as the colour image sequence with which they are associated. The depth maps can be stored as 8-bit per pixel grey values, where grey value 0 specifies the furthest position from the camera, and a grey level of 255 specifies the closest position to the camera. This depth map representation can be mapped to real, metric depth values. To support this, the grey levels are normalized into two main depth clipping plains. The near clipping plane Z_{near} (grey level 255), represents the smallest metric depth value Z. The far clipping plane Z_{far} (grey level 0), represents the largest metric depth value Z. In case of linear quantization of depth, the intermediate depth values can be calculated using the following equation:

$$Z = Z_{far} + v\left(\frac{Z_{near} - Z_{far}}{255}\right) \text{ with } v \in [0, \ldots, 255] \tag{1.1}$$

where v specifies the respective grey level value.

An example colour-plus-depth representation can be seen in Fig. 1.8.

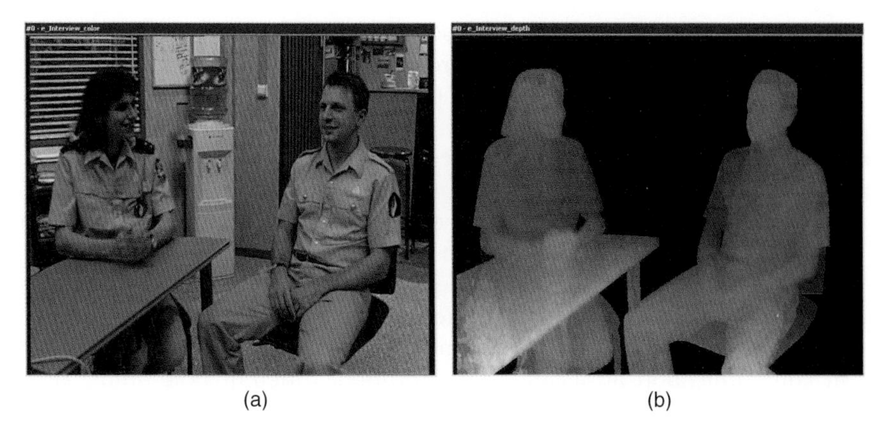

(a) (b)

Figure 1.8 Example colour-plus-depth video sequence commonly used by video coding experts for testing video compression algorithms, called Interview, produced in the course of ATTEST Project (a) Colour sequence. (b) Depth sequence (Reproduced with permission of the ATTEST Consortium)

Depth-Image-Based (DIBR) can be used to synthesize two views for the left and right eyes using colour image sequences, and the corresponding per-pixel depth map [37, 38]. This process requires two main steps: re-projection of original image point into 3-D space using depth information, and projection of the 3-D space points into the image planes of the left and right views.

The advantages of using colour-plus-depth map representation of stereoscopic video compared to video generated with a stereo camera pair can be summarized as follows:

- 3D rendering can be optimized for different stereoscopic displays and viewing scenarios to yield a disparity that is comfortable to the eye.
- Head-Motion Parallax (HMP) may be supported, which provides an additional 3D depth clue. This format also partially overcomes the viewing angle limitation of traditional stereoscopic camera set-ups.
- Most depth information does not have high frequency components. Thus, the depth sequence can be efficiently compressed with existing compression standards [39], and will require only limited space and bandwidth compared to that required by the colour image sequence.
- Photo-metrical asymmetries (e.g. in terms of brightness, contrast or colour) between the left and the right eyes, will be eliminated. Thus, the associated eyestrain problems will be avoided.
- Depth can be used in 3D post production (e.g. fine tuning of depth to eliminate stereoscopic artefacts that may occur during filming).

However, there are a number of drawbacks associated with this representation. The disadvantages and possible solutions are as follows:

- The quality of the rendered stereoscopic views depends on the accuracy of the per-pixel depth values. Therefore, the effects of compression and transmission of depth maps on the perceived quality need to be carefully considered (see Chapter 6).
- Objects that are visible in the rendered left and right views may be occluded in the original view. An example of this is shown in Figure 1.9. This phenomenon is also known as exposure and dis-occlusion in Computer Graphics (CG) [38]. This effect can be minimized using Layered Depth Video (LDV), where more than one pair of colour-plus-depth sequences is transmitted depending on the requirements of the expected quality. However, this approach demands more storage and bandwidth to be used in communication applications. In addition, different hole-filling algorithms (e.g. linear interpolation of foreground and background colour, background colour extrapolation, mirroring of background colour information) can be utilized to recover the occluded areas of the original image sequence [39]. Moreover, pre-processing/smoothing of depth maps (e.g. use of a Gaussian filter) will avoid this occlusion problem. However, this approach will lead to some distortions of the rendered 3D video scene.
- Certain atmospheric effects (e.g. fog, smoke), and semi-transparent objects are often poorly handled with this approach.

1.2.3 Multi-View Video

Multi-View Video (MVV) is a format required to support many promising 3D video applications, such as FVV and holographic displays (rendered as stereoscopic video in the simplest case, or multi-view auto-stereoscopic video in more advanced scenarios). Section 2.2.3 describes the camera arrays used

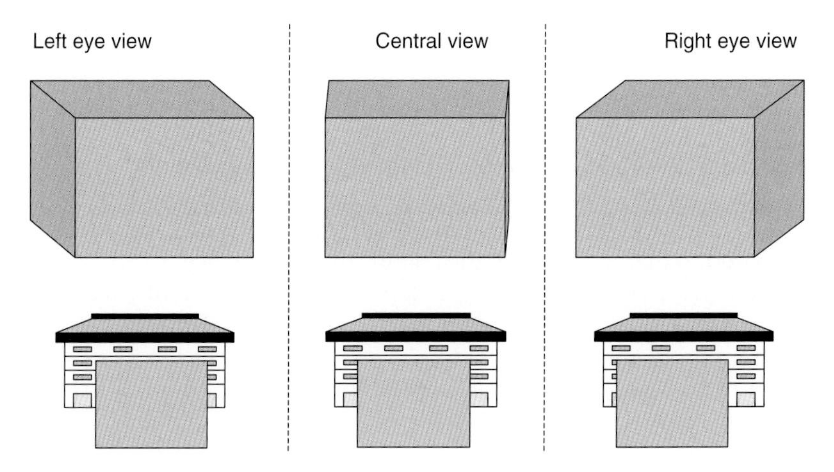

Figure 1.9 Illustration of occlusion problems with the colour-plus-depth format, where only a central viewpoint is stored, rather than the left and right views

to capture MVV in detail. The arrays provide several calibrated viewpoints of the same 3D scene, captured simultaneously. Multi-view video application, unlike conventional video applications (e.g. 2D HDTV or stereoscopic 3DTV), promise to offer a high level of interaction between the user and the content. FVV allows the user to navigate throughout the video scene.

Following extensive investigations into 3D video (mainly within the MPEG 3DV Ad Hoc group), multi-view video was widely recognized as a powerful video format and the need for a standardization activity for the compression of multi-view videos was identified. The results of these standardization activities are discussed in more detail in Section 3.3.2.

1.2.4 Multi-View Plus Depth Video

The colour-plus-depth format introduced in Section 1.2.2 provides very limited free-viewpoint functionality. The head motion parallax in the colour-plus-depth format can adjust the rendered video within a very narrow navigation range [40]. The multi-view video explained in Section 1.2.3 can theoretically provide a much larger scene navigation range. However, rendering using only colour views means that depth estimation must be performed at the receiver. Depth estimation errors can cause significant visual artefacts in the rendered video. Therefore, multi-view video itself cannot always provide good quality free-viewpoint rendering, particularly if the inter-camera baselines are large.

Multi-View plus Depth (MVD) is an extension of both the multi-view video format and the video-plus-depth format. It consists of N colour texture viewpoints and N corresponding per-pixel depth map sequences. The inclusion of depth extends the free-viewpoint navigation range [41, 42], as well as smoothness of scene navigation. Using the MVD format is it possible to render intermediate virtual viewpoints at any position, provided the spatial position vectors (rotational and translational matrices) and camera parameters of the target virtual viewpoint are present. MVD is a very powerful scene representation format enabling full feature 3DTV supporting multiple simultaneous viewers. Figure 1.10 depicts the multi-view plus depth representation, and shows the synthesis of arbitrary viewpoints with this representation.

Multi-view plus depth does not necessarily carry excessive additional transmission overhead compared to multi-view video, as the multi-view depth map sequences can be treated as colour texture videos and can be efficiently compressed using multi-view coding tools. These multi-view coding tools are described in Section 3.3.3. Extensive coding results can be found in [40] and [41].

1.2.5 Layered Depth Video

The concept of Layered Depth Image (LDI) or Layered Depth Video (LDV) was introduced in [43], and can facilitate efficient image-based 3D scene rendering. A layered depth image represents a scene using an array of pixels

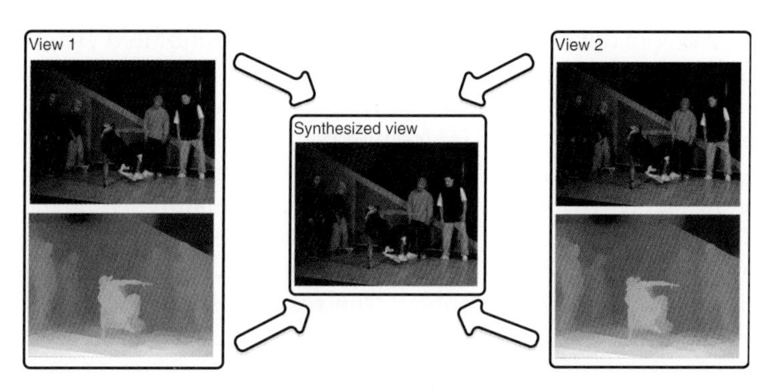

Figure 1.10 Multi-view plus depth format video allows synthesis of virtual views using colour and depth information from nearby viewpoints (Reproduced with permission of Microsoft)

viewed from a single camera position [44]. Each pixel in the array consists of colour, luminance, depth, and other data that assists 3D scene rendering. According to [44], three key characteristics of LDV are:

- It contains multiple layers at each pixel position.
- The distribution of pixels in the back layer is sparse.
- Each pixel has multiple attribute values.

As shown in Figure 1.11, intersecting points between the rays emanating from a reference camera, and an object are stored [44]. The information stored for each intersecting point consists of colour and depth information. The conceptual diagram of a layered depth image is depicted in Figure 1.11 [43, 44]. In Figure 1.11, the intersecting points closest to the reference camera form the first layer of the LDV. The second nearest are used in the second layer, and so on. Consequently, each Layered Depth Pixel (LDP) has a different number of Depth Pixels (DPs), each of which contain colour, depth, and auxiliary information for reconstruction of arbitrary viewpoints of the 3D scene [44]. For the specific example shown in Figure 1.11, LDP 4 has four layers, which contain data pertaining to the intersecting points between Ray D and the object. A single LDI is created per time instant (i.e. the LDI at time *t* is composed of the images and depth map values belonging to the same time instant from all different views).

Conventional video codecs, such as MPEG-4 AVC, are optimized to remove temporal and spatial redundancies, which are common for most 2D video sequences. However, the layers of LDV do not exhibit the same characteristics, in terms of spatial and temporal correlation. Conventional video coding tools are inefficient when coding the layers of an LDV. Therefore, although LDV is an excellent format for rendering, the lack of an efficient compression scheme is likely to hold it back.

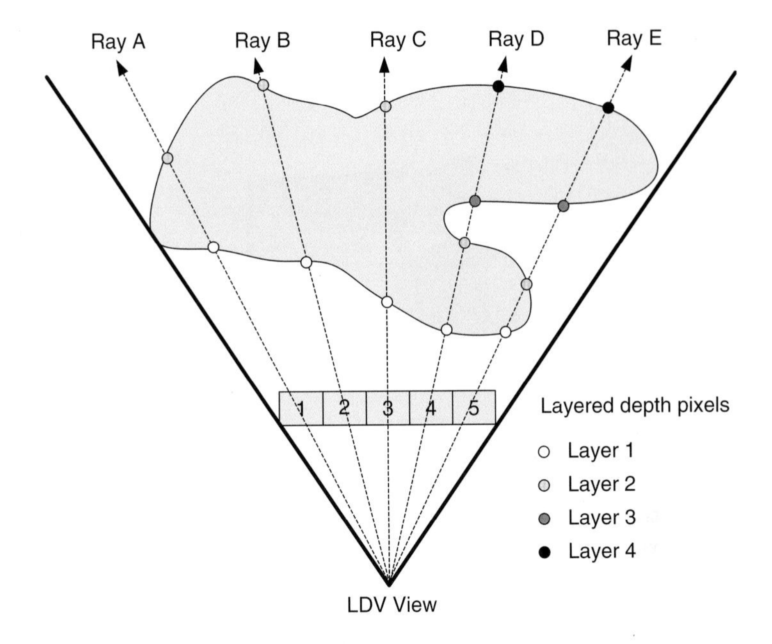

Figure 1.11 Concept of layered depth video

1.3 3D Video Application Scenarios

Although this book focuses mainly on 3DTV, the techniques and technologies described in this book are relevant to a range of application scenarios. This section describes some of the most interesting applications, including broadcast, mobile 3DTV, 3D video streaming, immersive video-conferencing, and remote applications (e.g. remote surgery, control of robots).

1.3.1 3DTV Broadcast Systems

Broadcast remains the most efficient way of delivering high quality video content to large numbers of people. In Europe, the digital television standards are being revised to enable support for 3DTV channels. Frame compatible format video is already supported by the various Digital Video Broadcasting (DVB) standards, while work is under way to enable support for service compatible video. On 1 October 2010, British Sky Broadcasting launched a 3D channel that could be decoded using their standard High Definition (HD) DVB-S2 set-top box. Compatibility with the existing technology is made possible by the use of frame compatible 3D, where the two views are downsampled in the horizontal direction, so that they can both be placed within one HD frame.

Figure 1.12 A Sony 3DTV, along with a consumer-level stereoscopic camcorder (Courtesy of Sony)

Unfortunately, the range of content currently available on 3DTV channels is somewhat limited. Much of the existing content is made up of 3D Hollywood movies. However, filming of live sports events has also been performed using stereoscopic camera rigs.

3D broadcasts have also taken place in other countries, including Japan, the USA, and Australia. As more cost-effective 3DTV sets are produced by mainstream manufacturers (such as the set shown in Figure 1.12), broadcasters are likely to introduce new premium 3DTV channels to increase their revenues. An interesting point to note about 3DTV services is that the success of 3D movies may be a significant driver of 3DTV sales. While the movie industry may have started 3D productions to keep people away from their televisions, the end result could be the successful establishment of 3D services for the home. In the 1950s and 1980s, this would have been unthinkable due to the costs. But good quality 3D televisions prices are dropping to prices low enough for the average consumer.

1.3.2 Mobile 3DTV

One of the original ideas of stereoscopic cinema was that it would provide a greater sense of realism and immersion. It may therefore appear to be a strange choice to try and introduce 3D services on to small mobile devices. Also, the use of 3D glasses in many mobile scenarios is likely to be inconvenient for users of the service. However, because mobile devices are typically viewed by an individual, rather than by a group of people, they are uniquely suited to many auto-stereoscopic approaches (see Section 5.4.4 for a more detailed description of this technology), which often require that the viewer positions themselves within a 'sweet spot' to view the video in 3D.

Commercial mobile devices with 3D displays are already starting to find their way to the market. Although not a mobile phone, Nintendo's 3D DS portable games platform is an interesting example of a small device with a 3D display. Manufacturers such as Samsung and Nokia have already begun to produce experimental mobile phones with 3D capability. In addition to this, research funded by the European Union has brought together coalitions of industry and academia to work on the following two projects:

- *3D PHONE* – examines various topics relevant to future 3D mobile phones, such as auto-stereoscopic displays, stereoscopic capture of video, 3D user interfaces, and compression of 3D video. The project partners are Bilkent University, Fraunhofer Heinrich-Hertz-Institute (HHI), Holografika, TAT (The Astonishing Tribe), Telefonica and University of Helsinki.
- *MOBILE 3DTV* – investigates how to deliver 3DTV services to a mobile hone with an auto-stereoscopic display, over DVB-H. Special attention is paid in the project to quality issues. The project partners are Tampere University of Technology, Technische Universitt Ilmenau, Middle East Technical University, Fraunhofer HHI, Multimedia Solutions Ltd and Tamlink Innovation-Research-Development Ltd.

Both of the above-named projects have successfully demonstrated 3D mobile phone technology, and have carried out user trials showing strong user acceptance.

1.3.3 3D Video on Demand

3D Video on Demand involves the delivery of video over Internet Protocol (IP)-based networks, when requested by the user. In the United Kingdom, Video on Demand has become popular, and has been branded for the public using names such as iPlayer, and Catch-Up TV. Many television sets are now being produced that are capable of being connected to the Internet to view VoD. The Consumer Electronics Show (CES) in 2011 featured a large number of so-called 'connected TVs', and Forrester Research predicts that [45]:

> Internet connected TVs will continue their steady penetration into consumers' homes, in large part due to retailers' commitment to only sell connected TVs in the future.

Forrester estimates that 58 per cent of all TVs sold will be 'connected' by 2015. Of course, if these connected TVs are also capable of displaying 3D, then the extension of these 2D video services to 3D is relatively trivial. Service providers can either use the frame compatible formats currently in use for

European DVB broadcasts, or frame sequential service compatible format (see Section 1.2). Problems with 3D VoD are likely to arise, when more views are required to support multi-view displays, which is clearly more demanding in terms of network bandwidth requirements. These transmission issues are discussed in more detail in Section 4.3.2.

1.3.4 3D Immersive Video-Conferencing

A number of companies have already produced high-end video-conferencing facilities, which attempt to provide greater sense of the remote participants being in the same room. Hewlett Packard and Cisco are two of the main proponents of such high-end facilities. Both companies offer systems that exploit High Definition video-conferencing, rendered on multiple large screens. Cicso's Telepresence 3000 system is shown in Figure 1.13, which demonstrates the basic configuration of such high end systems. People within the room are made to feel that the remote participants are present in the same room.

The introduction of 3D to such high-end video-conferencing systems would seem like a natural step forward in terms of immersion. However, there are some issues that currently limit the effectiveness of 3D for video-conferencing. These issues become apparent when we consider the currently available 3D display technology:

- *Passive or active stereo displays* – require that glasses are used. This may cause problems for participants to interact effectively, as it may inhibit eye contact with remote meeting participants.
- *Auto-stereoscopic displays* – typically feature sweet spots. Although some displays can be created with multiple sweet spots, viewers movement is

Figure 1.13 Cisco's Telepresence 3000 system, featuring high definition video-conferencing (Courtesy of Cisco Systems Inc.)

still heavily constrained, making it difficult to interact naturally with both remote participants and colleagues in the same room. There are also likely to be significant complexities in setting up multiple 3D displays, so that the sweet spots all converge on particular seating spaces.

The answers to these problems will likely either arise come from improving auto-stereoscopic displays, or from image processing techniques that are capable of making 3D glasses invisible to remote participants. However, this last image processing solution does not solve the problems with interacting with people in the same room.

1.3.5 Remote Applications

The medical community has conducted a number of studies into the use of 3D video for surgery. Stereoscopic video can be used for surgery that requires cameras (e.g. remote surgery, or keyhole surgery). 3D video can provide more accurate perception of depth, allowing the surgeons to make more accurate incisions. This in turn can lead to fewer complications arising from surgery, and faster recovery times.

1.4 Motivation

It is clear from the history of 3D video (see Section 1.1), that while public popularity of the technology has waxed and waned, development has continued. The repeated revivals point to real public interest in 3D technology. Previous failures have been caused by significant quality problems in the production and projection of 3D movies, and a lack of good affordable 3D technology for the home.

While significant improvements in camera and projection technology have been made over the years, as well as the introduction of affordable 3DTVs for the home, one of the most important developments has been the introduction of digital video processing technology. Many of the problems associated with previous 3D booms can be put down to production and projection problems that are difficult to spot and to fix with the naked eye. Digital technology allows analysis and correction of stereoscopic errors to sub-pixel accuracy. Examples of the benefits that can be made by digital technology are as follows:

- *Colour correction* – differences in contrast, and colour saturation between views may be precisely adjusted to ensure that stereoscopic views match (see Section 2.3.1). This reduces eye fatigue issues.
- *Rectification* – any problems with camera viewpoints can be corrected, as described in Section 2.3.1. An example of such a problem is when one camera, from a stereoscopic pair, points slightly upwards or downwards compared to the other.

- *Extraction of depth* – can be achieved using the kind of algorithms described in Section 2.3.2. Depth allows better free-viewpoint rendering, and also fine-tuning of disparities between viewpoints so that the content can be optimized to viewing position and screen size.

There are therefore many reasons to believe that the current revival in 3D is one that will prove more enduring than previous booms.

1.5 Overview of the Book

This book is split into seven chapters, most of which can be directly related to blocks in the 3DTV chain shown in Figure 1.14. Following this introduction, the book works its way through the 3DTV chain, starting with the capture of 3D video in *Chapter 2*. This chapter discusses capture of the various different formats for 3D video (see Section 1.2), and the camera technology currently available to perform the capture. Following capture, some image processing steps are often required. First the videos must be rectified and colour matched to prevent visual fatigue and to enhance the efficiency of multi-view video compression. Finally, depth must be extracted. Approaches for depth extraction are discussed in Section 2.3.2.

The captured and post-processed video is incredibly large, and therefore compression, as discussed in *Chapter 3*, is essential. The basics of video compression are briefly covered, before looking specifically into compression of the formats described in Section 1.2. Existing standards are covered, and the chapter also discusses current and future standardization efforts.

Chapter 4 examines approaches for transmitting 3D video. Transmission of stereoscopic video over broadcast or IP-based networks can be achieved using relatively straightforward extensions of existing 2D transmission schemes. The real problems occur when multi-view or holographic display support is required. In such cases, multiple HD views must be transported, placing a significant strain on communication networks. Some holographic displays require between eight and 64 views, which requires a bandwidth greater than can be provided by existing networks.

After delivery and decompression of the video, rendering and display must be performed. *Chapter 5* examines a variety of 3D rendering approaches.

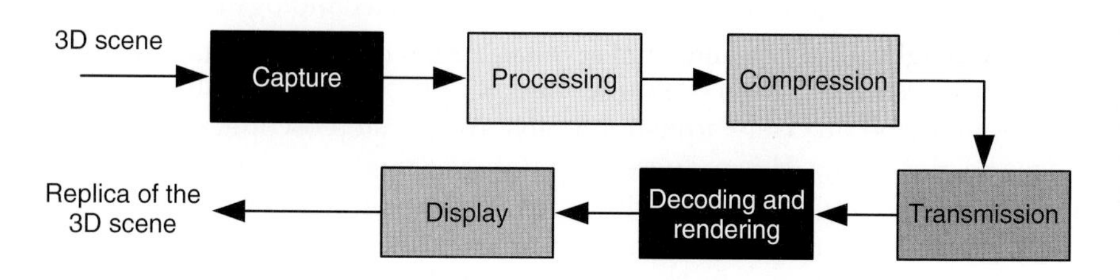

Figure 1.14 Basic building blocks for 3DTV

In addition, several adaptation types for 3D and multi-view video are also outlined here, such as network adaptation, context adaptation and adaptation to users preferences. Finally, the principles behind existing stereoscopic displays and more advanced displays technologies, which may become popular in the future (such as light-field and holographic displays), are looked at in this chapter.

Chapter 6 discusses quality and human perception of 3D video. This has become an increasingly important topic for 3D video, and understanding of these issues requires discussion of the physiology and psychological elements of the Human Visual System (HVS). The chapter discusses how to carry out subjective viewing tests in order to obtain reliable quality results. It also discusses methods for evaluating 3D video using objective metrics. This kind of approach is significantly faster than carrying out subjective testing, but may not always produce reliable results.

The final section of the book, *Chapter 7*, summarizes the book, and provides pointers to which improvements are necessary, and which improvements are likely to be made, in future years. It also makes a brief assessment concerning the continued success of 3D video applications.

References

[1] Lipton, L. (1982) *Foundations of the Stereoscopic Cinema: A Study in Depth*, Van Nostrand Reinhold, Amsterdam.

[2] Hayes, R.M. (1998) *3-D Movies: A History and Filmography of Stereoscopic Cinema*, McFarland & Company, Jefferson, NC.

[3] Wheatstone, C. (1838) 'Contributions to the physiology of vision. Part the first. On some remarkable, and hitherto unobserved, phenomena of binocular vision', *Philosophical Transactions of the Royal Society of London*, **128**, 371–394, June.

[4] Rollmann, W. (1853) 'Notiz zur Stereoskopie', *Annalen der Physik*, **165** (6), 350–351.

[5] Coe, B. (1969) 'William Friese Greene and the origins of cinematography', *Screen*, **10** (2), 25–41.

[6] Lynde, D. (1915) 'Stereoscopic pictures screened', *The Moving Picture World*, p. 2072, 26 June.

[7] Screen (1922) 'Vivid pictures startle', *The New York Times*, 28 December.

[8] Herapath, W.B. (1852) 'XXVI. On the optical properties of a newly-discovered salt of quinine, which crystalline substance possesses the power of polarizing a ray of light, like tourmaline, and at certain angles of rotation of depolarizing it, like selenite', *Philosophical Magazine Series 4*, **3** (17), 161–173.

[9] John Anderton, 'Method by which pictures projected upon screens by magic lanterns are seen in relief', Patent number 542321, July 1895.

[10] John Anderton, 'An improved method or system by means of which pictures projected upon a screen from an optical or magic lanterns or lanterns are seen in relief or with stereoscopic effect', Patent number GB189801835 (A), November 1898.

[11] Edwin Herbert Land, 'Polarizing refracting bodies', Patent number US1918848 (A), July 1933.
[12] *NY Times* (1936) 'New "glass" cuts glare of light; aid to movies and science seen; inventor shows polarizing substance which, with spectacles of same material, dims brilliance of auto lamps – may be used in beauty treatments.', January.
[13] *Encyclopedia Britannica* (2011) 'House Un-American Activities Committee (HUAC)' available at: www.britannica.uk.
[14] Land, E.H. (1940) 'Vectographs: Images in terms of vectorial inequality and their application in three-dimensional representation', *Journal of Optical Society of America*, **30** (6), 230–238.
[15] Edwin Herbert Land, 'Light-polarizing image in full color', Patent number 2289714, July 1942.
[16] Robert V. Bernier, 'Three-dimensional adapter for motion-picture projectors', Patent number US2478891 (A), August 1949.
[17] Robert V. Bernier, 'Stereoscopic camera', Patent number CA787118 (A), June 1968.
[18] *BOXOFFICE Magazine* (1982) 'UACs stereospace', November.
[19] International 3D Society (2010) '3D movie fans expand box office says international 3D society study', March.
[20] Berthier, A. (1896) 'Images stéréoscopiques de grand format', *Cosmos*, **34**, 205–210.
[21] Ives, F.E. (1902) 'A novel stereogram', *Journal of the Franklin Institute*, **153** (1), 51–52.
[22] Frederick E. Ives, 'Parallax stereogram and process of making same', Patent number 725567, April 1903.
[23] Lippmann, M.G. (1908) 'Épreuves réversibles. Photographies intégrales', *Comptes Rendus de l'Académie des Sciences*, **9** (146), 446–451.
[24] Herbert E. Ives, 'Projection of stereoscopic pictures', Patent number 1883290, October 1932.
[25] Clarence W. Kanolt, 'Stereoscopic picture', Patent number 2140702, December 1938.
[26] D. Gabor, 'System of photography and projection in relief', Patent number 2351032, June 1944.
[27] F. Savoye, 'Stereoscopic motion-picture', Patent number 2441674, May 1948.
[28] E.H.V. Noaillon, 'Art of making cinematographic projections', Patent number 1772782, August 1930.
[29] R.B. Collender, 'Three dimensional unaided viewing method and apparatus', Patent number 3178720, April 1965.
[30] John Logie Baird, 'Improvements in or relating to stereoscopic pictures', Patent number GB292365 (A), June 1928.
[31] Tiltman, R.F. (1928) 'How "stereoscopic" television is shown', *Radio News*, November.
[32] James Frank Butterfield, Willard Stanton Alger, and Daniel Leslie Symmes, 'Stereoscopic television system', Patent number GB2114395 (A), August 1983.
[33] Sand, R. (1992) '3-DTV-a review of recent and current developments', in *IEE Colloquium on Stereoscopic Television*, London, October, p. 1, IRT stereo system.

[34] Terry D. Beard, 'Low differential 3-D viewer glasses and method', Patent number EP0325019 (A1), July 1989.

[35] Imaizumi, H. and Luthra, A. (2002) 'Stereoscopic video compression standard: MPEG-2 multiview profile', in *Three-Dimensional Television, Video, and Display Technologies*, Springer-Verlag, New York, pp. 169–181.

[36] Yuyama, I. and Okui, M. (2002) 'Stereoscopic HDTV', in *Three-Dimensional Television, Video, and Display Technologies*, Springer-Verlag, New York, pp. 3–34.

[37] McMillan, L. (1997) 'An image-based approach to three-dimensional computer graphics', PhD thesis, University of North Carolina at Chapel Hill, Chapel Hill, NC.

[38] Mark, W.R. (1999) 'Post-rendering 3D image warping: visibility, reconstruction, and performance for depth-image warping', PhD thesis, University of North Carolina at Chapel Hill, Chapel Hill, NC, April.

[39] Fehn, C. (2004) 'Depth-image-based rendering (DIBR), compression and transmission for a new approach on 3DTV', *Proceedings of SPIE Conference on Stereoscopic Displays and Virtual Reality Systems XI*, **5291**, 93–104.

[40] Merkle, P., Smolic, A., Müller, K. and Wiegand, T. (2007) 'Multi-view video plus depth representation and coding', *IEEE International Conference on Image Processing*, **1**, 201–204.

[41] Kauff, P., Atzpadina, N., Fehna, C., Miller, M., Schreer, O., Smolic, A. and Tanger, R. (2007) 'Depth map creation and image-based rendering for advanced 3DTV services providing interoperability and scalability', *Signal Processing: Image Communication*, **22** (2), 217–234.

[42] Zitnick, C.L., Kang, S.B., Uyttendaele, M., Winder, S. and Szeliski, R. (2004) 'High quality video view interpolation using a layered representation', *ACM Transactions on Graphics*, **23**, 600–608.

[43] Shade, J., Gortler, S.J., He, L.W. and Szeliski, R. (1998) 'Layered depth images', *SIGGRAPH*, **July**, 231–242.

[44] Yoon, S. and Ho, Y.S. (2007) 'Multiple color and depth video coding using a hierarchical representation', *IEEE Transactions on Circuits and Systems for Video Technology*, special issue on multi-view video coding and 3DTV, **17** (11), 1450–1460.

[45] Warman, M. (2011) 'CES 2011: tablets and televisions set to dominate a connected show', *The Daily Telegraph*, 5 January.

2

Capture and Processing

The capture and preparation of content are one of the most important parts of the 3DTV chain. As described in Section 1.1.3, many of the movies produced in the 1950s had significant problems that led to significant discomfort for viewers. These problems included vertical parallaxes between the two stereoscopic views, too great a physical distance between the two views, and also inconsistencies in processing of the two views after filming.

All of these problems mean that 3D video requires very careful and meticulous planning and set-up. The cameras need to be physically positioned in a way to prevent discomfort for viewers. The cameras must also be carefully configured so that their settings match precisely (e.g. focus, shutter speed). The introduction of digital technology means that there is greater tolerance to imperfections in the camera configuration. Digital post-processing techniques allow views to be aligned, and colours to be matched. However, to ensure the best quality, it is advisable to get as close to the optimum configuration as possible during filming.

This chapter explores how stereoscopic and multi-view video can be captured using specially configured rigs. In addition, it looks at the post-processing techniques that can be used to align viewpoints, and correct colour variations across views. Finally, depth extraction techniques are considered, which are essential for generating content in the Multi-view Video plus depth (MVD) format.

2.1 3D Scene Representation Formats and Techniques

Before considering the technologies and approaches for 3D video capture, we should review the scene representation formats and consider what kind of data is required for each format. Table 2.1 summarizes this information and

3DTV: Processing and Transmission of 3D Video Signals, First Edition.
Anil Fernando, Stewart T. Worrall and Erhan Ekmekcioğlu.

Table 2.1 Summary of 3D video formats and their respective requirements placed on video capture systems

Format name	Data types required	Potential capture system technologies
Frame compatible stereoscopic video	Two colour video views, which should feature a small horizontal disparity	Two separate, synchronized cameras may be used, or alternatively specially manufactured cameras featuring two lenses and one or two sensors
Service compatible stereoscopic video	The same input data as service compatible is required. The only difference is that two separate full resolution fames are needed	The same set-up as frame compatible may be used, but two separate sensors are ideally required for full resolution service compatible video
Colour-plus-depth	Colour video is required from a single camera. Depth images are used to represent the distance from the camera	Depth can be captured using multiple synchronized colour cameras. Depth can then be extracted using image processing approaches. Alternatively, Time-of-Flight cameras can be used which output depth data directly
Layered depth video	Multiple colour and depth views are needed, where each layer contains colour plus depth data	Capture of real scenes (rather than generation using CGI), requires carefully arranged arrays of cameras so that all occlusions can be eliminated
Multi-view video	Multiple camera views are needed, along with data about the relative distances between the cameras in the array	Systems using arrays of synchronized cameras are typically used. The cameras are usually carefully positioned and aligned using special rigs
Multi-view video plus depth	Similar to multi-view video, but with added depth information	The systems are usually very similar to the multi-view case, but can also feature. Time-of-Flight cameras

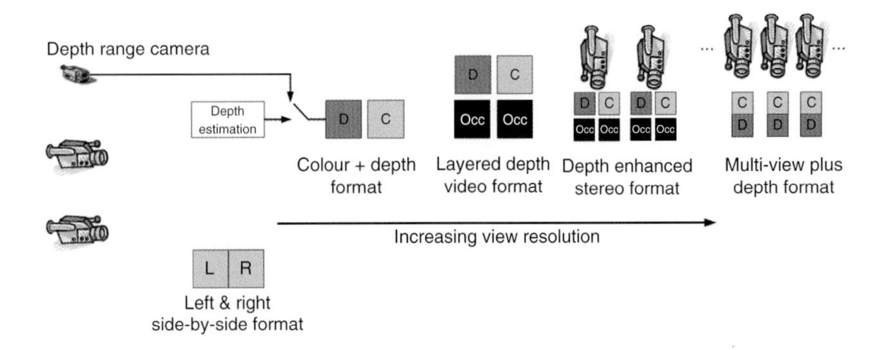

Figure 2.1 Graphical visualization of 3D video formats

also covers the basics of how data might be captured for each format, whereas Figure 2.1 depicts the increasing 3D resolution from simpler formats. The two most common types of information needed are colour video and depth information (see Section 1.2.2).

Colour video is usually achieved using multiple standard 2D cameras fixed onto special rigs. The depth data may be acquired in a variety of ways. This is commonly done by estimating the depth from the multiple 2D camera views (see Section 2.3.2), or by using a special type of camera, called Time-of-Flight (ToF), as described in Section 2.2.1.

The actual methods used to generate 3D data depend on whether the data is to be generated from real-world scenes, or from Computer Generated Imagery (CGI). This chapter covers capture of 3D data for representing real-world scenes, rather than CGI. Some of the principles for working with CGI are the same as with real-world data. However, it is much easier to accurately place the multiple cameras, and many packages also have an option to generate highly accurate depth information. Examples of such packages include the free Blender software [1], where on-line tutorials can be found describing stereoscopic movie rendering, and generation of depth information.

2.2 3D Video Capturing Techniques

In this section, we consider methods of producing the 3D video data. The camera technologies used in 3D capture systems are described, and methods for acquiring both stereoscopic and multi-view video are examined. The discussions also include examples of 3D video capturing, drawing on experience from work on EU-funded research projects, MUSCADE and DIOMEDES. These projects were funded by the European Union, and involved the acquisition of multi-view video as part of the overall research programme.

2.2.1 Camera Technologies

Although post-processing can be performed to correct issues that arise during video capture, selecting the right cameras can reduce the amount of post-processing required. This in turn results in better quality 3D content.

There is a limited selection of cameras manufactured specifically for 3D content capture. Broadcast quality 3D cameras are even more difficult to find. As a result, much 3D video has been produced using standard 2D cameras, mounted on special rigs. These cameras must be precisely synchronized and calibrated to ensure accurate 3D rendering for the view, without causing visual fatigue.

However, recently a number of consumer electronics manufacturers have introduced 3D-capable camcorders to their project ranges. Panasonic, for example, have introduced a new lens for their camcorders, which is compatible with some of their existing 2D camcorders. The lens projects the two stereoscopic views onto a single sensor, which means that the camcorder captures the video in side-by-side format. This is convenient for many consumer 3D displays, which accept side-by-side video as an input. The problem with this arrangement is that the resolution is compromised. Sony and JVC have both introduced competing products, which feature two separate sensors, enabling High Definition capture of stereoscopic video. To make it possible for consumers to capture 3D video easily, the camcorders try to estimate and vary the amount of disparity between the two views, depending on the scene content and the focal length used. Although they are quite effective products for consumers, they do not allow the amount of control over capture needed for professional quality 3D filming.

2.2.1.1 Key Requirements for Standard 2D Cameras

Before considering the requirements for the cameras, it is important to know what will be done with the captured video. For 3D capture, it may be necessary to use the 2D video to estimate depth (see Section 2.3.2), compress multiple camera views jointly, and to use the video to render intermediate viewpoints. Let us consider what each of these implies:

- *Estimation of depth* – the cameras should have high resolution and good quality.
- *Joint compression of multiple views* – the captured video views should be similar in terms of brightness, contrast, and colour saturation, so that the difference between views is minimized. This provides better compression efficiency when compressed with MVC (see Section 3.3.2).
- *Rendering intermediate viewpoints* – as above, the video should be similar in terms of brightness, contrast, and colour saturation. This will prevent noticeable changes in the picture when the viewpoint is changed.

This means that, in addition to having high quality cameras, it is important to have a high degree of manual control over the camera settings. This enables fine adjustment during equipment set-up, to produce the closest possible matches between captured viewpoints.

2.2.1.2 Time-of-Flight Cameras

Time-of-Flight (ToF) cameras are special types of sensors, which aim to capture three-dimensional images at high frame rates. The acquired images comprise per-pixel depth intensity values. They are advantageous in the field of 3D video research, due to their compact size (no need for additional mobile parts) and light-weight structure. This has been made possible by the rapid advances in the field of micro-optics and micro-technologies in general. Figure 2.2 presents sample shots from some of the Time-of-Flight cameras on the market. The acquired intensity images resolutions are depicted next to each camera. In the scope of 3D video capturing and 3DTV systems, they are used along with a series of 2D cameras, as described previously, in order to capture the depth maps and generate the colour-plus-depth format for encoding. Nevertheless, the current level of technology still imposes some limitations on the capability of depth map acquisition:

- *Limitation on the resolution of the captured range images* – Current Time-of-Flight cameras on the market can provide relatively lower resolution

Mesa Imaging AG, SR3000 (QCIF - 176 × 144)

Mesa Imaging AG, SR4000 (QCIF - 176 × 144)

PMD Technologies, 19k (160 × 120)

PMD Technologies, CamCube3.0 (200 × 200)

Figure 2.2 Sample Time-of-Flight camera products that are currently on the market

depth image sequences, ranging from 64×48 to 204×204. Thus, it has always been crucial to apply post-processing on them to obtain higher resolution depth maps, equal to the resolution of the relevant camera viewpoint.

- *Limitation on the quality of the captured image sequences* – The noises in the acquired range images are another source of limitation. Noises in the depth values still cannot be completely removed after calibration.

Time-of-Flight range cameras have widely replaced 3D depth acquisition systems that are based on a camera and lens. Camera-based systems are rather old-fashioned and work by estimating the depth fields in a scene based on the variations in the instantaneous focus. Similarly, the other methods used are depth estimation for motion and from changes in the geometric structures. The disadvantage of using camera-based systems in computing the depth information has been the fact that in order to remove the frequently observed ambiguities, spatial and temporal correspondences have to be thoroughly worked out over a sequence of captured images. Accordingly this results in increased computational complexity and processing costs.

On the other hand, laser-based depth acquisition systems have been able to capture the depth field more accurately and with a higher range precision than camera-based systems. Nevertheless, they are bulky and more voluminous than Time-of-Flight cameras and are more difficult to operate (e.g. they usually need to be mounted on a unit that can be adjusted using tilt and pan movements, in order to maximize the field of view for a particular scene).

In the scope of 3DTV and corresponding content acquisition systems, ToF cameras have usually been used in combination with supplementary high-resolution 2D cameras, in order to aid real-time high quality multi-view depth estimation because ToF cameras produce registered dense-depth and colour intensity image sequences. Thus, after diffusing the low resolution depth map values to other supplementary cameras (via a camera coordinate transformation using the extrinsic information of both cameras) and upscaling the transformed depth pixels intensities using the appropriate image processing techniques (i.e. exploiting the inter-camera structural correspondences), higher resolution and higher precision multi-view depth map sequences can be generated. Interested readers are referred to [2], to find out more about a practical application of such a hybrid model in estimating dense multi-view depth maps.

2.2.2 Stereoscopic Video Capture

The stereoscopic video can render the left and right views for both eyes to produce a 3D impression. The binocular disparity, which represents the dissimilarity in views due to the relative location of each eye, is exploited most, to help the human visual system to perceive the depth. Other physiological depth cues include convergence, motion parallax, chroma-stereopsis; and psychological depth cues include image size, perspective and occlusion.

As the simplest and most cost-effective way of creating stereoscopy, L-R stereoscopic video is captured by a pair of calibrated cameras. The shooting parameters can be determined once and applied accordingly during capture. The camera baseline distance (the distance between the optical axes of shooting cameras), the convergence distance (the distance of the cameras from the point where both optical axes intersect) and the focal lengths are among the shooting parameters. Accordingly, the horizontal disparity between two viewpoints and, hence, the degree of perceived depth in the scene are controlled. Two different camera configurations are possible: parallel camera configuration (the optical axes never intersect) and toed-in camera configuration (the optical axes intersect at the convergence distance). Both are schematically sketched in Figure 2.3.

The main point is to correctly align the images captured by both of the cameras by manual and also electronic means. Correct alignment means matching the cameras, focal lengths, white balance levels, and removing all vertical parallax existing between the cameras (i.e. making all the corresponding epipolar lines of both cameras parallel to each other). At present, the most professional solution is the use of two customized HD cameras mounted on a concrete rig (cameras can be placed side-by-side or they can be mounted on a mirror rig too) and the most widely available output is an independent pair of time-synchronized image sequences, providing the left-eye and right-eye views.

Several companies have produced customized stereoscopic video acquisition rigs. Most of these systems provide the cameramen with the flexibility of removing the vertical alignment of the cameras by manual and electronic fine adjustment, and changing the inter-camera distance, depth of focus, convergence of the cameras (i.e. the amount of deviation between the angles of the focal lines of each camera) as well as other system parameters employed by stereographs to achieve artistic ends. Figure 2.4 depicts a list

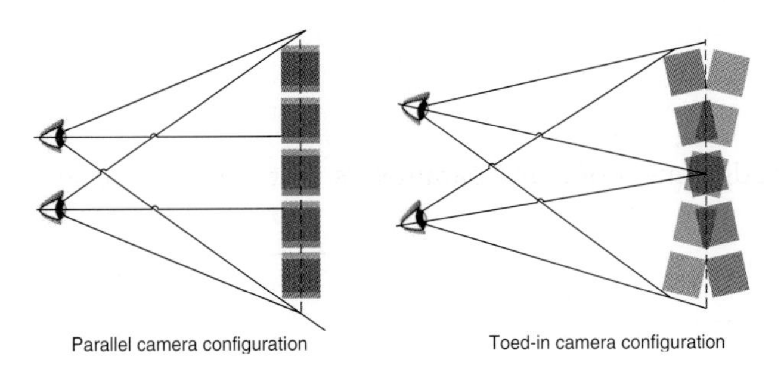

Figure 2.3 Dual (stereoscopic) camera configurations
Source: http://binocularity.org

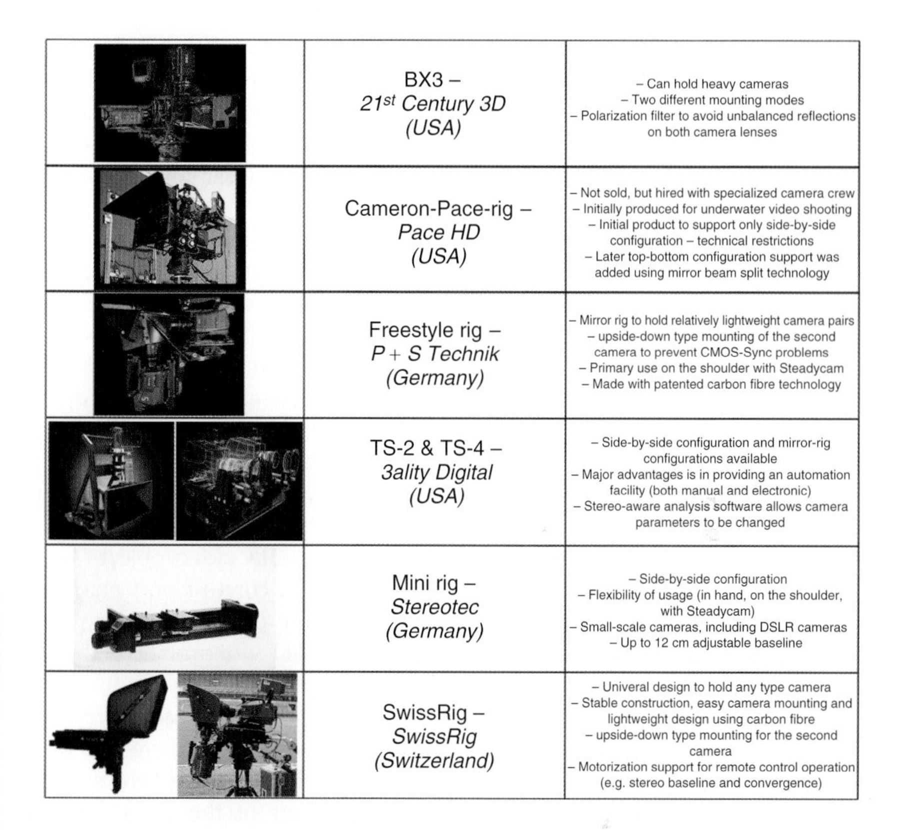

	BX3 – *21st Century 3D (USA)*	– Can hold heavy cameras – Two different mounting modes – Polarization filter to avoid unbalanced reflections on both camera lenses
	Cameron-Pace-rig – *Pace HD (USA)*	– Not sold, but hired with specialized camera crew – Initially produced for underwater video shooting – Initial product to support only side-by-side configuration – technical restrictions – Later top-bottom configuration support was added using mirror beam split technology
	Freestyle rig – *P + S Technik (Germany)*	– Mirror rig to hold relatively lightweight camera pairs – upside-down type mounting of the second camera to prevent CMOS-Sync problems – Primary use on the shoulder with Steadycam – Made with patented carbon fibre technology
	TS-2 & TS-4 – *3ality Digital (USA)*	– Side-by-side configuration and mirror-rig configurations available – Major advantages is in providing an automation facility (both manual and electronic) – Stereo-aware analysis software allows camera parameters to be changed
	Mini rig – *Stereotec (Germany)*	– Side-by-side configuration – Flexibility of usage (in hand, on the shoulder, with Steadycam) – Small-scale cameras, including DSLR cameras – Up to 12 cm adjustable baseline
	SwissRig – *SwissRig (Switzerland)*	– Univeral design to hold any type camera – Stable construction, easy camera mounting and lightweight design using carbon fibre – upside-down type mounting for the second camera – Motorization support for remote control operation (e.g. stereo baseline and convergence)

Figure 2.4 Some off-the-shelf stereoscopic 3D rig products

of commercial/specialized stereoscopic video capture rigs by stereoscopic camera manufacturers and their representative pictures.

2.2.2.1 Key Requirements of Stereoscopic Video Capture Systems

Creating a stereoscopic video content, as mentioned, is not as easy as just capturing two time-synchronized 2D video sequences with an approximate eye-distance between each camera. Previous experience has shown that the improper adjustment of many parameters that are involved in stereoscopic video acquisition may lead to unacceptably poor 3D video quality that causes eyestrain and headache [3].

Among the most basic two parameters of any stereoscopic video capture rig is the adjustment of the inter-axial distance and the amount of convergence introduced between the cameras. It should be noted that 3D visualization is an illusion of creating the feeling of the real 3D scene on the screen plane. Thus, every 3D video object should be exposed to the viewers within the comfort zone, which is usually measured by the ratio of the 3D screen's width to the average distance of the viewer from the screen. The challenge is to adjust these two parameters, such that all video objects within the reconstructed 3D visualization fit within the comfortable depth limits with respect to the screen, and yet look realistic as close as possible to the actual scene that is shot.

While the inter-axial distance between the cameras usually contributes to the overall disparity between the shot stereoscopic video pair, hence the global depth, the amount of introduced convergence changes the relative depth positioning of each video object with respect to the screen plane. The improper adjustment of the camera convergence usually results in binocular "keystone"-type distortion that is known to be caused by attempting to project the video onto the screen plane at a certain angle. This distortion increases with an increase in the distance between the object and the camera, decreasing convergence distance and decreasing focal distance [4]. Furthermore, if the convergence between the cameras is not properly adjusted, the viewer can perceive wrong distances between 3D video objects' surfaces and the screen plane, where their shapes are corrupt and changed from the natural representation [4]. Accordingly, the projected scene is cut into only a sparse set of depth planes and most 3D video objects appear unnaturally flat.

Apart from the afore-mentioned settings, there are some other types of well-known distortions/artefacts (not necessarily connected to stereoscopic video acquisition systems), unless the stereoscopic video capture system is well adjusted. These include: blurriness in either of the pairs (by the loss of focus in either lenses), barrel distortions (through decreased magnification with increasing distance from the optical axis), spatial aliasing (by insufficient sampling of the acquired 3D video signal), motion blur (appears on fast-moving objects where the camera is unable to map them sharply on to the sensor), unbalanced photometry between both cameras (mismatches in colour saturation levels, differences in brightness, contrast, and gamma values).

Having explained the types of commonly observed stereoscopic video artefacts and their causes, one can appreciate that if the rigging and the calibration of the stereoscopic camera pair are done accurately (i.e. accurate matching of electronic and optical camera parameters and the adjustment of the inter-camera baseline according to the depth structure of the captured scene), the captured stereoscopic video content can be as free-of-artefacts as possible. However, note that in order to bring the captured content up to entertainment and broadcast quality, usually a set of post-production tools need to be applied. More details on the post-production tools will be provided in Section 2.3.

2.2.2.2 Automation of Stereoscopic Video Capture Systems

It is a pretty time-consuming task to properly adjust all settings of a stereoscopic video capture system depending on the characteristics of the scene to be shot, and usually requires skilled staff (e.g. stereographs). However, currently there is an increasing demand for efficient 3D production tools to assist the camera crew in order to ease the time-consuming steps of stereoscopic capture monitoring on the shooting set. An automatic scene analysis system is able to estimate the relative poses of both cameras in order to allow the optimal camera alignment and adjustment of lens settings in real time. In addition, it is able to detect the relative positions of video objects in the scene (via a number of segmentation techniques) to derive the optimal baseline distance and warn the cameraman of stereoscopic window violation, i.e. when a particular video object appears on the edge of one camera view but is not visible in the other camera view. Not many such systems are available today in the market, or are being developed.

Within the scope of two of the recent EU-funded projects, Content Generation and Delivery for 3D Television (3D4YOU) [5] and Multimedia Scalable 3D for Europe (MUSCADE) [6], such an automatic stereoscopic video analyser tool has been developed that aims to operate in real time to monitor the quality of the captured stereoscopic video in terms of stereoscopic cues. The latter project (MUSCADE) has aimed to extend the stereoscopic video analyser to a multi-camera analyser (consisting of four cameras, the details of which will be discussed in Section 2.2.3) that is completely image-based. An overview of the workflow of the developed stereoscopic video analysis tool is depicted in Figure 2.5.

The first step of the deployed stereoscopic video analysis tool consists of downsampling the luminance images, finding points of interest using a feature-tracking algorithm and matching the point correspondences between

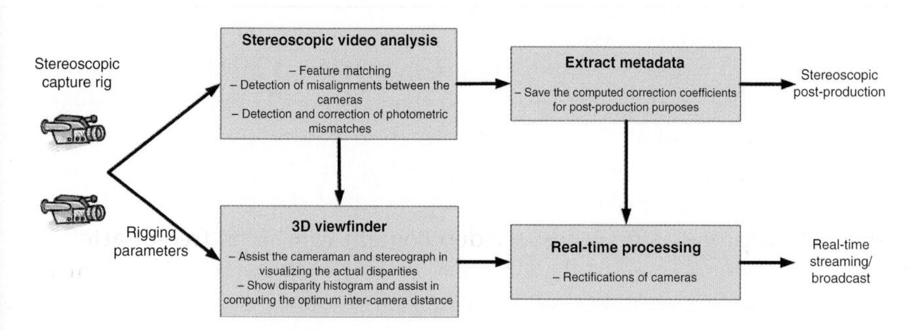

Figure 2.5 The workflow of the stereoscopic video analyser described in (3)

the two stereoscopic video frames [3]. In order to identify robust matches, to estimate the position of both cameras, and to compute the parameters for the correction of camera misalignments, and keystone distortions, the constraints of epipolar geometry are taken into consideration.

Similarly, photometric parameters are analysed to detect any mismatches. Accordingly, all computed geometric and photometric correction parameters are either stored as a metadata file for offline post-production purposes, or are directly applied in the adjustment of the stereoscopic camera rig in real time. These adjustments involve steering the lens control, changing the electronic camera settings and camera positioning in the case of motorized lenses and rig, and interfacing with the camera signal processing. In addition to these processes, the developed stereoscopic video analysis tool also facilitates an intuitive graphical user interface and is able to show the cameraman and the stereograph the histogram of the current disparity levels in the shot 3D video, and thus the current clipping depth planes of the scene. The cameraman is free to adjust the convergence and the inter-axial distance between the cameras accordingly. However, the analysis tool is itself able to compute the optimum inter-axial distance, once it has fixed the other rigging parameters and cameras' focal lengths, based on the constraints of achieving the 3D video visualization within the comfortable viewing zone. Interested readers can refer to [3] and to [7] in order to obtain more information about the working principles of the described real-time stereoscopic video analyser tool.

2.2.2.3 2D-to-3D Conversion for Stereoscopic Video Generation

After the spread of the three-dimensional visualization systems, another core research area has been converting the existing (previously recorded) programs into stereoscopic 3D. The term 2D-to-3D conversion refers to using monocular depth cues obtained from a 2D video sequence to generate the equivalent 3D video sequence.

The 2D-to-3D video conversion has been motivated mostly by two types of scenarios. The main one is to convert the existing 2D video archives into a new 3D version of it. The other motivation is about shooting/creating a completely original 3D video that is shot in 2D (e.g. using a single video camera). This could be due to the fact that the cost of maintenance and operation of a 3D capturing system would be more than that of a well-established 2D capturing systems. The conversion process primarily necessitates the segmentation of the 2D image sequence. For each segmented video object, the relative depth is computed using two-dimensional visual cues. It involves locating the occlusion areas then (i.e. the video regions which are visible to one of the cameras, but is not visible to the other camera) and concealing them using the appropriate pixel interpolation techniques and using the texture information of the surrounding segments.

The 2D-to-3D video conversion can be done in real time or offline, depending mainly on the computational complexity and temporal memory usage of the applied processing steps. Nevertheless, it should be noted that the 2D video segmentation process has been the bottleneck preventing the online conversion. One of the main shortcomings of the 2D-to-3D conversion processes is that the depth values are assigned to each segmented video object, rather than each pixel. Thus, pixels categorized under the same segment are assigned the same depth value. Unfortunately, this would make the video objects rendered after the conversion process look flat (e.g. the stereoscopic artefact called the "cardboard effect") and hence unrealistic.

Another fact is that despite the ever increasing research efforts applied to fully automatic real-time 2D-to-3D video conversion systems (e.g. especially for use on several TV sets and/or set-top boxes to give the users the chance of watching the programs in 3D on the fly), it is really quite difficult to optimize the 3D visual experience thoroughly. Furthermore, it could also be an off-putting factor for the users of 3DTV, since they cannot tolerate long-term eye-strain and unrealistic visual distortions, as well as headaches, as a result of poor 3D visual experience. On the other hand, efforts to semi-automatically convert existing pre-recorded 2D movies and programmes offline, with careful adjustment of disparities across objects throughout the scene depending on the motion and object locations offer a more promising way to produce high quality 3D content.

Offline 2D-to-3D video conversion would usually follow the following sequence of processes: (1) estimation of a depth map using a single frame of 2D video; (2) synthesizing the second view by using semi-automatic in-painting techniques for filling in the holes; and (3) sending the results to the temporally neighbouring video frames via a set of tracking techniques [8]. Readers who are interested in learning the recent 3DTV-related activities, especially across in European research institutes, universities and companies, are referred to [8].

2.2.3 Multi-View Video Capture

The interest in the research community in capturing high quality multi-view content has increased thanks to the recent advances in multi-view coding (more details are given in Chapter 3 of this book) and streaming techniques, paving the way to realize immersive 3D media distribution. Nevertheless, building a multi-view capture rig (usually with more than two cameras) has its own challenges that include multi-camera calibration, timely synchronization of each camera, rectification, and auxiliary information extraction (e.g. accurate depth maps).

Similar to stereoscopic video capturing rigs, multi-view capturing systems should also be constructed with several constraints based on the actual application area. Obviously, multi-view video is needed to reconstruct a denser 3D visualization of a natural scene, resulting in a more realistic

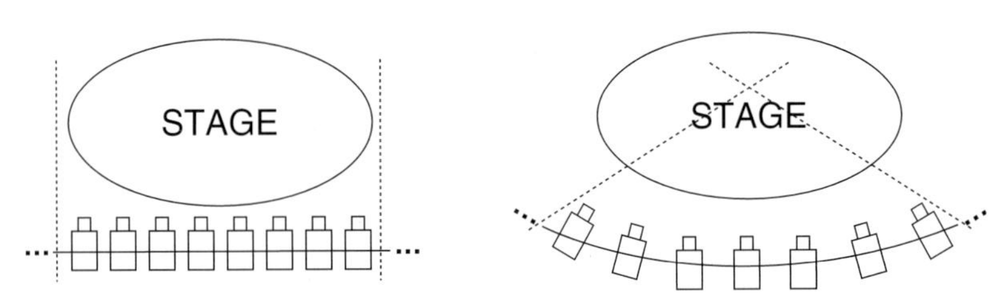

Figure 2.6 Sample one-dimensional multi-camera arrangement drawings (left: linear arrangement, right: arc arrangement)

perception that just a stereoscopic video. Auto-stereoscopic, multi-view and more sophisticated light-field and holographic displays, which will be explained in more detail in Chapter 5, are the main focus of multi-view video applications.

On the other hand, stereoscopic and/or conventional 2D displays are still suitable to use with the multi-view video content in the context of free-viewpoint video applications. In other words, multi-view video is useful in the sense that the users are free to change their viewpoint in the 3D scene. Details of the multi-view and multi-view plus depth video formats were presented in Chapter 1 and an overview was provided in Section 2.1.

A one-dimensional multi-camera arrangement is the most convenient way of achieving multi-view video, because current stereoscopic and multi-view display technologies are suitable to provide horizontal parallax, i.e. they work on the assumption that the head movement is in the horizontal direction mostly. Thus, it is not necessary to capture the visual information on the top and the bottom of the 3D scene objects. Two popular one-dimensional camera arrangement types are the linear camera arrangement (the cameras' optical axes are parallel) and the arc camera arrangement (the cameras' optical axes are convergent). They are schematically depicted in Figure 2.6. The key requirements of a practical multi-view capture system can be summarized as:

- *The Cameras should be well calibrated* – The intrinsic parameters of each camera should be well computed. The Cameras' focal lengths and aspect ratios should be identical. Ideally, all the cameras used within the multi-camera array should be of the same brand and model.
- *The rotation matrices should be balanced* – For linear 1D camera arrays, where the optical axes of all cameras are parallel to each other; the 3×3 rotation matrix of each camera should be identical.
- *The convergence points of cameras should coincide as much as possible* – For arc-type multi-camera arrays, one should try to adjust the cameras rotation angles, so that the optical axes of each camera should intersect as much at the same point as possible. Note that it is very difficult to do a manual

adjustment to create a common convergence point. A post-production stage is necessary to transform the acquired camera views.

- *Vertical displacement between cameras optical axes should be minimized* – Vertical displacement (or vertical disparities) between the cameras would result in serious visual artefacts in the rendered multi-view video on the screen, because human eyes can only naturally perceive horizontal parallax. Thus, all the corresponding epipolar lines within all the camera views should coincide. Note that it is very difficult to ensure the removal of all such vertical disparities within a multi-camera rig, where a post-rectification stage is crucial, which will be explained in more detail in Section 2.3.1.
- *The inter-axial distances between the cameras should be adjusted properly* – It is important that the multi-camera rig is customized depending on the shot of the 3D scene. It is essential to avoid too much coincidence between the Field Of View (FOV) of neighbouring cameras. This depends on the proximity of the multi-camera rig to the shot scene and thus the applied focal lengths. At the same time, the views of the outer cameras in the multi-view array should coincide with the outer view of the targeted displays (i.e. all captured visual information should fit into the FOV of the display).
- *It should be possible to extract accurate depth/disparity information from the captured multi-view video* – Accurate and dense disparity/depth information per each camera makes it possible to synthesize a high-quality intermediate viewpoint, either within the actual multi-camera baseline, or outside the baseline of the multi-camera system. Accordingly, it is much easier to address the display output requirements of multi-view displays and light-field displays, which usually comprise outputting many more viewpoints than what is originally delivered to them (i.e. synthesizing M views out of N reconstructed camera views, where $M \gg N$). The depth extraction process will be explained in more detail in Section 2.3.2.
- *The setting of all cameras should be matched as best as possible in terms of contrast, brightness, gamut, primaries, etc.* – A post-processing stage is usually necessary to compensate for all such inter-camera differences (e.g. colour correction that will be described in Section 2.3.1).

Three recent EU-funded projects that are concentrated on an end-to-end implementation of modern 3DTV chain have also considered several different multi-view capture architectures. In the following, a basic overview of the deployed multi-view video capture systems is provided.

2.2.3.1 4-Camera Multi-View Rig of MUSCADE Project

The multi-view acquisition system in MUSCADE is based on four identical high-definition video cameras. An overview of the deployed acquisition system is depicted in Figure 2.7. Two of the cameras, the inner pair, form a regular stereoscopic video pair. This way, it is targeted to directly support

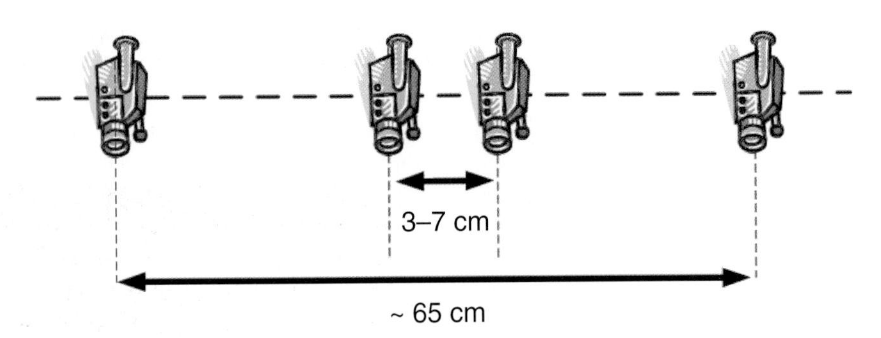

Figure 2.7 Overview of the MUSCADE 4-cam multi-view video acquisition system (7)

conventional stereoscopic TV systems in a multi-layer 3D video service architecture. Hence, backwards compatibility is achieved. The inner stereoscopic camera pair is mounted on a conventional mirror rig. The usual range of baseline is between 3 cm and 7 cm, as depicted in Figure 2.7.

The outer pair represents a wide baseline stereo set-up with a camera distance of approximately 65 cm or more. The main purpose of the outer cameras is to provide additional visual information for accurate depth/disparity map extraction, where the extracted depth/disparity information is used in the synthesis process of intermediate virtual camera views.

2.2.3.2 8-Camera Multi-View Rig of DIOMEDES Project

The 8-view multi-camera rig has been constructed using eight high-end Thomson® Viper® HD cameras in a 1D parallel, side-by-side arrangement. The cameras are mounted in a fixed position on a long steel rig. Due to the chunky size of the digital cinema quality cameras, the cameras have been squeezed as much as their size allowed. On the other hand, the physical restriction of the cameras has been compensated for to a certain extent by a relatively large range of permissible focal length: from 4.7 mm to 52 mm. The cameras are evenly spaced with an approximate distance of around 21.5 cm between each.

The vertical disparity between the cameras is initially minimized as much as possible by means of manual control of the camera pan and tilt mechanisms. The electronic viewfinders placed on each camera, as well as an external monitor, to which the output of all eight cameras are connected, are used in the manual adjustment stage for vertical disparity minimization, before proceeding to post-processing. Colour adjustment of each camera is also controlled remotely before the actual shooting that minimizes the photometric differences between the cameras.

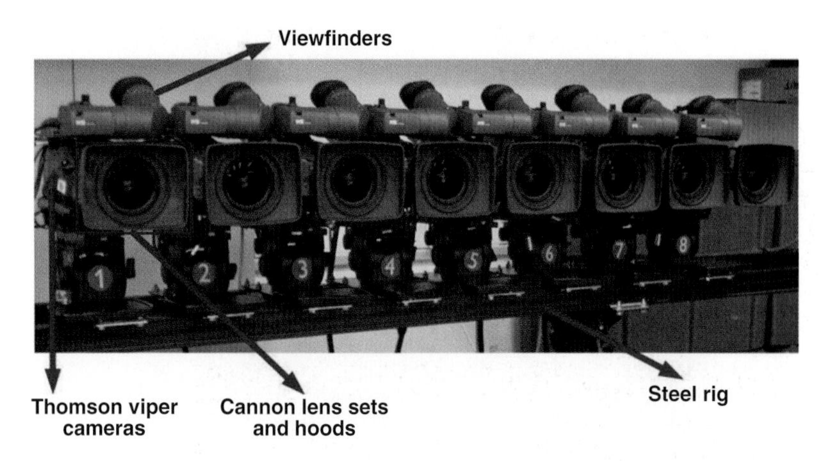

Figure 2.8 Overview of the DIOMEDES 8-cam multi-view video acquisition system (9)

In order to simultaneously record the captured uncompressed HD sequences from eight cameras, four DVS® ProntoXway® video recorders are used, each of which accepts input from two of the cameras. All the eight cameras have been genlocked via an external synchronizer. The constructed multi-camera rig is presented in Figure 2.8. All videos are stored on an NTFS Windows® file system, with each video frame written to a separate image file. The video frames are stored as a sequence of 10-bit DPX images.

2.2.3.3 4-Camera Multi-View Rig of ROMEO Project

For multi-view video acquisition, a multi-view capture system has been constructed using four STEMMER AVT GT1910C cameras, which is depicted in Figure 2.9. Unlike the static multi-view calibration system used in the DIOMEDES Project, the ROMEO multi-view capture rig is transportable with a fixed camera focal length of 25 mm. As in the MUSCADE multi-view rig, the inter-camera distances are adjustable, thanks to freely moving camera trays. However, the inner stereoscopic pair is formed using a side-by-side camera configuration instead of using a mirror rig. With this multi-view acquisition system, the tunable disparity between the centre camera pair and the outer cameras obtains different disparities and then depth budgets for the same scene. In this particular project, several different inter-camera distance arrangement schemes were tested and compared to each other in terms of the created FOV of the reconstructed 3D scene on a set of targeted 3D displays. In the compared arrangement schemes, the distance between the inner camera pair and the outer cameras was changed between 10 cm to 20 cm, while the baseline of the inner camera pair is kept constant.

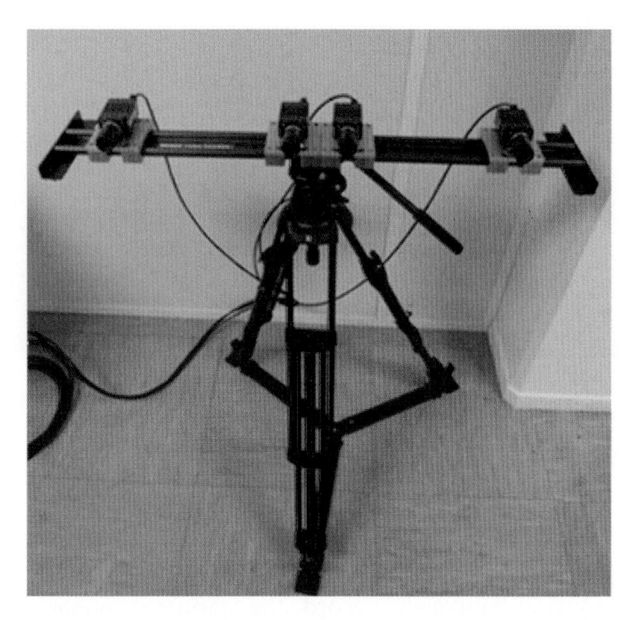

Figure 2.9 Overview of the ROMEO 4-cam multi-view video acquisition system (10)

2.2.4 Integral Imaging Capture

Integral imaging can capture accurate 3D colour image sequence using a single camera and it can offer pleasing 3D visualization experiences to a large audience at the same time, without leading to headaches and eyestrain in the long term. 3DTV services that are based on 3D integral imaging video technology, in which only a single camera is needed to capture the content, have the potential to become an attractive solution to service providers. The main reason is that this avoids the costly and time-consuming setting-up of multi-camera rigs, which most of the other modern 3D capture techniques employ.

Integral imaging is a technique that is capable of creating and encoding a true volume spatial optical model of the object scene in the form of a planar intensity distribution by using unique optical components [11, 12]. Since the integral imaging technique does not require coherent light sources, unlike holographic image capturing systems, it is usually more flexible and is used in live capturing events. The first person who coined the term integral 3D imaging was Lippmann [13] in 1908.

In order to capture an integral photograph, Gabriel Lippmann used an evenly spaced array of small lenses, which were closely packed together and were in contact with a photographic emulsion. Accordingly, each lens refracted the scene from slightly different angles to its neighbour lens, and thus, a scene was being captured from multiple viewing angles. When the photographic transparency is re-registered with the original recording array and illuminated by diffuse white light from the rear, the scene object is reconstructed in space by the intersection of ray bundles being emitted by each lens. The reconstructed video object is the integration of the pencil beams, which distinguishes integral imaging from holography. In the playback of the recorded 3D integral image, the reconstructed video object is inverted in depth. However, there have been optical and/or digital techniques to convert the inverted objects in depth to orthoscopic objects. Figure 2.10 depicts a graphical illustration of how a basic integral imaging system operates [14].

The main attraction of the integral imaging techniques currently on the market are its ability to easily support and deliver content to 3D displays without using glasses. Toshiba had already released its first 3DTV, featuring glasses-free multi-parallax technology in 2010, that is suited to integral imaging approach, where multiple users can watch a particular 3D scene from different viewing angles at the same time. The display produced by

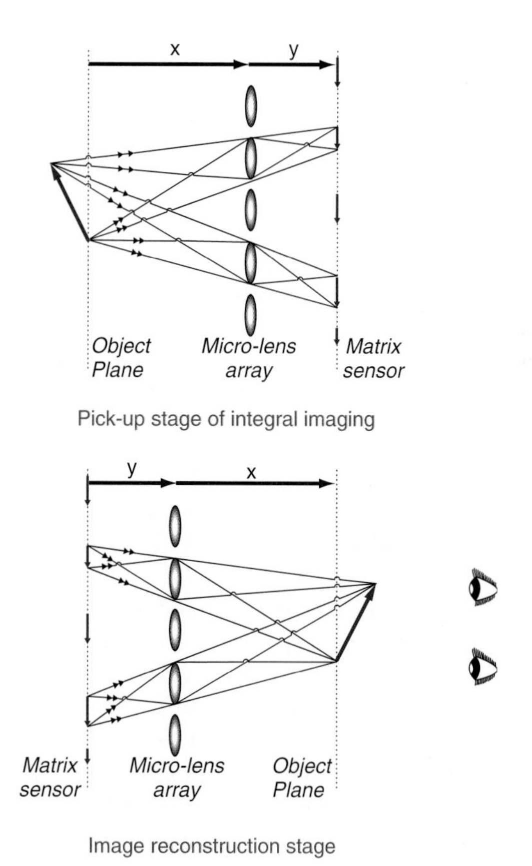

Figure 2.10 Capture (top) and display (bottom) stages of an integral imaging system (14)

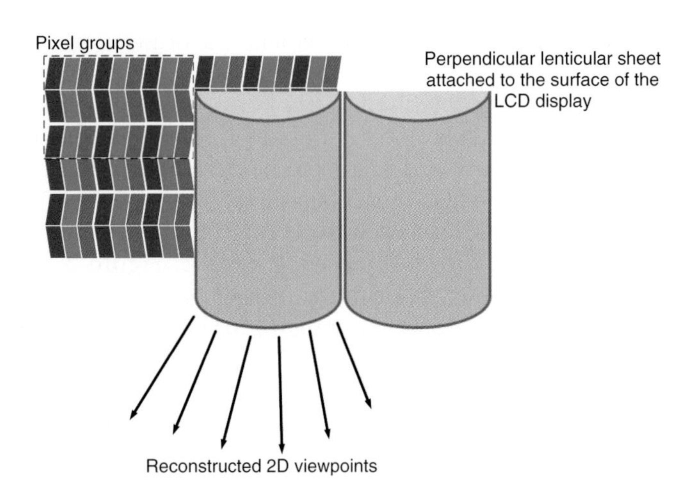

Figure 2.11 Toshiba's multi-parallax glasses-free 3DTV display system deploying integral imaging principles

Toshiba delivers nine parallax images that are refracted into nine different spots. In order to create these views, each pixel of the display panel is clustered into a group of nine, which then can display the same detail of the picture from nine slightly different viewing angles. These pixel clusters are covered with an orthogonal lens array to refract the generated pictures at the correct angle pattern. Figure 2.11 depicts this example.

Although integral imaging capture systems are convenient from the point of view of the capture set-up adjustment framework, it is not possible to capture a scene with a larger baseline than a carefully designed multi-view camera rig (i.e. the viewpoint navigation capability is very limited to the vicinity of the central viewpoint of the capturing camera).

2.3 3D Video Processing

The video streams captured by a rig of cameras should be processed prior to encoding and transmission, in order to balance the optical differences between different cameras, align them by transforming their coordinate systems and adapt the video streams to the external capturing conditions, such as colour distribution and exposure parameters. Most 3D video processing frameworks start with the calibration of the individual cameras within the multi-camera capture rig. In other words, each camera is calibrated separately first.

Camera calibration refers to the process that determines all the relevant camera intrinsic parameters that describe how an image is captured. Thus, it is possible to deduce the real-world coordinates of an object from its image coordinates. This includes all the parameters that determine the relation between the real-world spatial coordinates and the corresponding coordinates of the captured image. Similarly, the cameras are also jointly calibrated, where the relative positioning and rotational differences between the cameras are recorded. This calibration process corresponds to the extraction of the camera's extrinsic parameters. The intrinsic parameters model the defaults of each camera itself (i.e. lens distortion, focal length, etc.), whereas the extrinsic parameters denote the coordinate system transformation from the world coordinates to 3D camera coordinates, and vice versa.

Figure 2.12 depicts the extracted parameters in both calibration processes, and the corresponding flow diagram of the calibration process. The point positions on the test pattern, which is shown in Figure 2.12, are known beforehand and the goal of this manipulation is to find the intrinsic and extrinsic parameters, so that it is possible to obtain the best correspondence

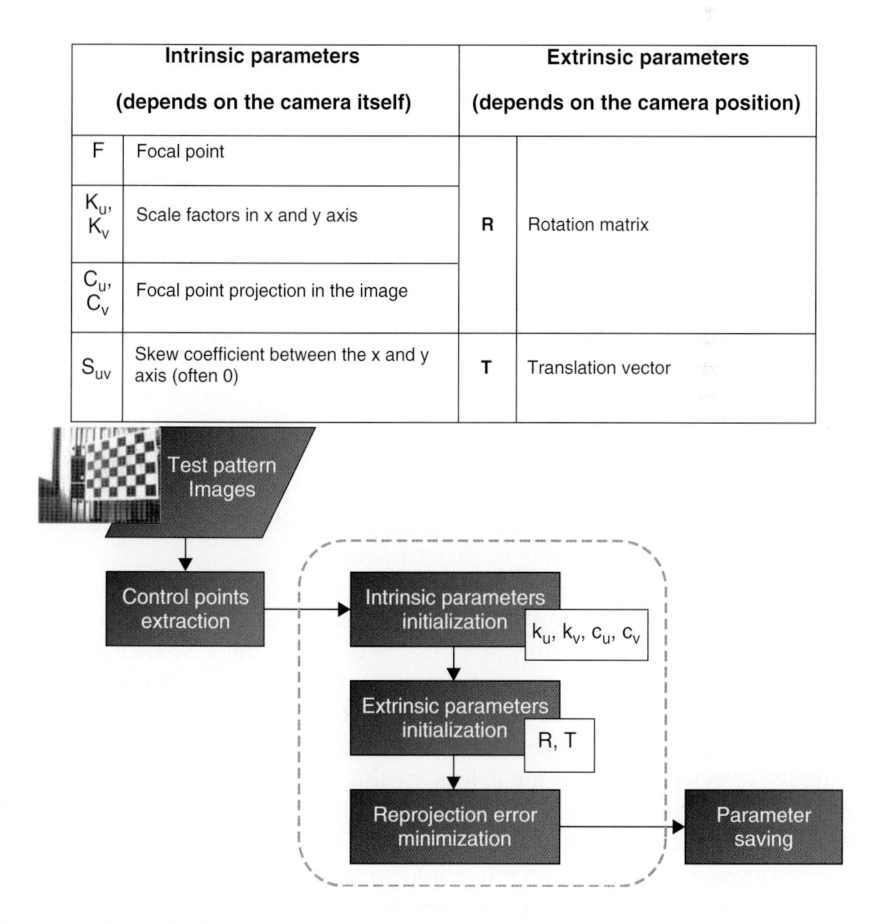

Intrinsic parameters (depends on the camera itself)		**Extrinsic parameters (depends on the camera position)**	
F	Focal point	**R**	Rotation matrix
K_u, K_v	Scale factors in x and y axis		
C_u, C_v	Focal point projection in the image		
S_{uv}	Skew coefficient between the x and y axis (often 0)	**T**	Translation vector

Figure 2.12 Camera calibration parameters and the workflow

between the control point and their projection. Once all the parameters are computed, it is possible to eliminate distortions in the image captured.

2.3.1 Rectification and Colour Correction

Recorded and calibrated multi-view image sequences should be converted into a lossless format to carry out the actual post-processing steps. Multi-view rectification is usually applied as the first step of a 3D multi-view post-capture workflow. Rectification is necessary in order to align the epipolar lines of each camera view, as well as to compensate for the slight focal length differences existing between the cameras. Accordingly, if the vertical disparities are removed completely by rectification, multi-view depth estimation search range can effectively be degraded to a single dimension (horizontal dimension only) in order to estimate the horizontal disparity and the dense stereo matching process becomes greatly simplified.

An epipolar rectification method has been described in [15], which is proposed for stereo image pair rectification. The described linear rectification algorithm accepts the projection matrices of the original cameras as input and computes a pair of rectifying projection matrices. The 3×4 projection matrix is the multiplication of the 3×3 camera intrinsics matrix (comprising the focal length, principal point and radial distortions) and the 3×4 camera extrinsics matrix (comprising the 3×3 rotation matrix and the 3×1 translation matrix in a side-by-side format). The original method suggests modifying the projection matrices of both camera views included in a stereoscopic rig. On the other hand, it is possible to modify the described method to account for the simultaneous rectification of more than two cameras. In other words, any two cameras selected from the N-camera ($N > 2$) multi-view rig should be rectified after the global rectification process. For this purpose, it is practical to design a batch, pair-wise rectification process, such that one of the cameras in each selected pair is common, and the projection matrix of that camera is not modified. On the other hand, the projection matrices of all other cameras are modified to have them rectified with the unaltered common camera. Eventually, it is possible to rectify each camera.

Multi-view colour correction is another essential step in the post-production workflow. Colour correction is important for many reasons. First, the dense stereo matching process to estimate the disparity is badly affected if there are colour differences in two cameras in the matched regions. Hence, the quality of disparity estimation is altered as a result of potentially wrong matches. Second, the multi-view video coding performance can be badly affected in terms of a loss in the rate-distortion performance, since the inter-view prediction scheme produces higher energy residuals in cases of luminance difference between the predicting and the predicted camera views. Third, the free-viewpoint synthesis process is also negatively affected if there are colour differences between the two source cameras (which are warped to the image coordinates of the target viewpoint) that need blending. 3D images look bad if the two source images have different colours. To overcome all such negative effects, the multi-view videos need colour correction before depth estimation and coding.

A histogram-based pre-filtering method had been described in [16] that is applied on the rectified multi-view camera images for colour correction. This method comprises the derivation of a mapping function that adapts the cumulative histogram of a distorted sequence to the cumulative histogram of a reference sequence. In the case of multiple cameras, all camera views should be adapted to a common reference camera view. The adaptation should be done for all R, G and B channels separately for every frame. Figure 2.14 shows a plot of the luminance histograms of the three sample images: the reference image used to correct the colour of the distorted image, the distorted image before colour correction, and the distorted image after colour correction. The reference image comes from the common camera view, whereas all other images in other camera views are regarded as distorted. Note that the bins in the histogram of the distorted image have been modified to match the distribution of the bins in the histogram of the reference image after colour correction. The colour differences are more significant especially at the peaks of the histogram, i.e. where the peaks of the reference and the distorted image do not align. From Figure 2.13 it is clear that most such peaks in the distorted

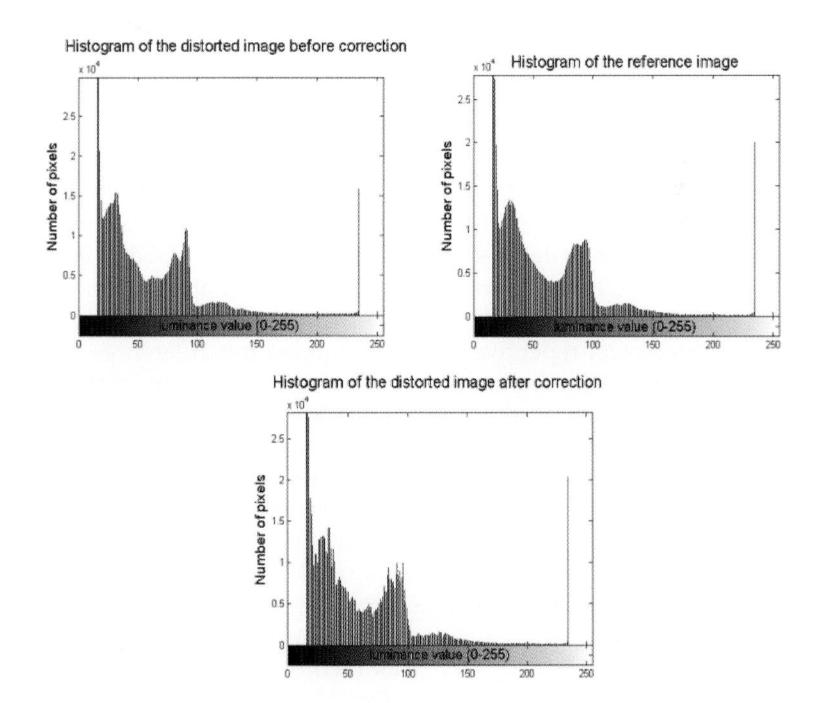

Figure 2.13 Luminance histograms of the reference image, the distorted image before and after colour correction

image that do not match the ones in the reference image are modified to get a closer match (e.g. peaks around luminance values = 30, 90).

2.3.2 Extraction of Range Images

Disparity map estimation/extraction consists of finding, for a given pixel in the reference image of any calibrated stereoscopic image pair, the corresponding pixel in the second image, so that the distance measured between them corresponds to the estimated disparity.

Knowledge of the disparity map can be used in the further extraction of information from stereo images. For instance, a disparity map is very useful in the calculation of depth, which represents the distance of the video object from the cameras. Disparity and depth, i.e. the distance from the cameras, are inversely correlated in such a way that, as the distance from the cameras decreases (closer video objects), the disparity increases. This enables the perception of depth in stereo images. Using geometry and algebra, the points that appear in the 2D stereo images can be mapped as coordinates in 3D space. Equation (2.1) outlines the algebraic relation between the disparity and the depth.

$$\text{Disparity} = x_l - x_r = (f \times T)/(Z \times t_{pixel}) \tag{2.1}$$

where:

f is the focal distance

T is the translational distance between the cameras of the stereoscopic pair

Z is the distance between the object and the cameras, i.e. the depth

t_{pixel} is the corresponding width of one pixel of the camera sensor

The process of accurate dense stereo depth extractions methods is complex and time-consuming, especially due to the image areas that include occlusion and at object boundaries, or fine structures, which would appear blurred in the depth maps. Lower repetitive textures, illumination differences and bad recording conditions can also make this task even more complicated. In the literature there are many different techniques applied to stereoscopic video pairs to extract the disparity/depth map information, all of which come at different expenses of the computational load. Techniques that rely on non-iterative processes tend to be faster than the ones that have multiple process loops, whereas iterative processes tend to enhance the sub-pixel disparity information by taking into consideration spatial/temporal and other types of correlations besides stereo matching cost.

An algorithm based on semi-global matching and mutual information method has been described in [17]. This method yields reasonably accurate dense disparity information and can provide relatively fast processing options subject to several software optimizations. The Stereo Global Block Matching (SGBM) method is based on block matching of mutual information and the approximation of a global 2-dimensional smoothness constraint by combining multiple 1-dimensional constraints. The method can be divided into three steps:

1. *Block matching cost calculation*: The aim of this step is to compute the luminance difference between two searched areas in the left and right images to find the corresponding pixels. In order to find them, the left image (or right image) is divided into blocks and each block is compared with a same size block in the right image (or left image) that is moved in a defined range (called "Horopter"), on the same line. A matching function computes the cost associated with each tested area. The cost is calculated as the absolute minimum difference of luminance intensities between the compared blocks. Several parameters are adjustable during this step (e.g. the size of the matching window, the size of the blocks, etc.) that can affect the accuracy of the cost calculation. All the cost values throughout the image pair are saved to be used in the next step.
2. *Cost aggregation*: The basic block matching cost calculation can yield ambiguous values and incorrect matches can produce a lower cost than the correct matches, due to a poor awareness of image content. Therefore, a further constraint is added in the second step that enforces a smoothness criterion by penalizing disparity value variations in the vicinity of neighbourhood. Thus, the cost aggregation is based on a penalty-reward scheme, and it progresses across eight directions originating from the block under concern.
3. *Disparity computation*: Following the cost aggregation stage in the previous step, the disparity map is computed by selecting for each pixel p the disparity d that corresponds to the minimum calculated cost. To avoid problems arising in the image regions with poor texture information, a secondary parameter called the "uniqueness constraint" is introduced. The incorporation of this parameter enforces the checking of the consistency across a region and adapting the estimated disparity accordingly.

The MUSCADE Project [6] (as previously mentioned in Section 2.2) has also dealt with the extraction of dense disparity information for four cameras in the multi-view rig, following the rectification and colour correction steps. The basis of the extraction of dense disparity information for a multi-camera set is pair-wise extraction of stereoscopic correspondences.

The stereo processing stage is based on using a matching algorithm called the Hybrid Recursive Matching (HRM) that is followed by an adaptive cross-trilateral median filtering (ACTMF) post-processing step to regularize the extracted disparity values. The initial values for the disparity are determined by applying the hybrid recursive block matching method that operates both in spatial and temporal (i.e. across successive video frames) directions to yield smooth and consistent estimates. The initial disparity value estimation process is applied in both directions, i.e. from right to left and from left to right to produce the disparity estimates in each camera. Both estimates are accordingly compared against each other for a consistency check and to produce a confidence kernel making use of the normalized cross-correlation.

In the successive adaptive tri-lateral median filtering stage, an adaptive filter is applied on the initially estimated disparity values for which the weighting coefficients are determined based on three different factors: the confidence kernel, the results from the colour texture analysis (i.e. the similarity in luminance intensity of the centre pixel and the weighted pixel) and the proximity (the distance from the centre pixel). Accordingly, at each current pixel position s, where the filter is centred, a disparity intensity value

Figure 2.14 Disparity estimation results using the mentioned HRM method (in the middle) and the successive ACTMF stage (at the bottom) for the left camera of ''Café'' MPEG test sequence by Gwangju Institute of Science and Technology (Reproduced with permission of Professor Yo-Sung Ho)

is determined by weighted median filtering using the neighbour pixels. The weighting coefficients are directly proportional to the confidence kernel value, so that the higher the reliability of a disparity estimate, the higher the chance of influencing the final estimate of the pixel in the centre of the filter. Figure 2.14 depicts an example of what the extracted disparity map looks like (as a grey-scale image) from the corresponding stereoscopic camera pair. The disparity map shown in the middle of Figure 2.14 represents the result after applying the first step (i.e. the hybrid recursive matching step), whereas the disparity map shown at the bottom represents the result after applying the adaptive median filtering based post-processing.

After the disparity is estimated for a pair of cameras inside the multi-camera rig and post-processed, the disparity values for the remaining outer cameras are found by back-projecting the estimated disparity values for the processed camera pair. It should be noted that for the remaining outer cameras, the same recursive stereoscopic matching stage with the subsequent post-processing stage is also applied, but the accuracy of these estimates is generally inversely proportional to the baseline between the tested cameras. In that particular research project, the baselines of the outer cameras were more than a stereoscopic baseline. Thus, since for the outer cameras the baseline is high, the estimated disparity map information is merged with the back-projected higher accuracy disparity estimates from the first pair that is tested (i.e. the inner pair with a stereoscopic baseline, please refer to Section 2.2.3).

References

[1] B. Foundation, 'Blender official website', June 2011.

[2] Kim, S.-Y., Koschan, A., Abidi, M.A., and Ho, Y-S. (2010) 'Three-dimensional video contents exploitation in depth camera-based hybrid camera system', in *High-Quality Visual Experience: Creation, Processing and Interactivity of High-Resolution and High-Dimensional Video Signals*, Springer, New York.

[3] Zilly, F., Muller, M., Eisert, P. and Kauff, P. (2010) 'The stereoscopic analyzer: an image-based assistance tool for stereo shooting and 3D production', in *IEEE International Conference on Image Processing (ICIP 2010)*, pp. 4029–4032.

[4] Meesters, M.J., IJsselsteijn, W.A. and Seuntiens, P. (2004) 'A survey of perceptual evaluations and requirements of three-dimensional TV', *IEEE Transactions on Circuits and Systems for Video Technology*, **14**, 381–391.

[5] http://www.3d4you.eu, 'EU ICT 3D 4 You Project.'

[6] http://www.muscade.eu, 'ICT MUSCADE Project (Multimedia Scalable 3D for Europe).'

[7] Zilly, F. and Malchow, W. (2012) '3D capture and post-production development phase 1', Tech. Rep., doc. D2.2.2, Multimedia Scalable 3D for Europe MUSCADE Integrated Project, February.

[8] Grau, O., Borel, T., Kauff, P., Smolic, A. and Tanger, R. (2011) '3D-Tv R and D activities in Europe', *IEEE Transactions on Broadcasting*, **57**, 408–420.

[9] Ekmekcioğlu, E. (2010) 'Report on 3D media capture and content preparation', Tech. Rep. D3.1, DIOMEDES EU ICT Project, September.

[10] Tizon, N. (2012) 'Report on 3D media capture and content preparation', Tech. Rep., D3.1 ROMEO EU ICT Project, May.

[11] Adedoyin, S., Fernando, W.A.C. and Aggoun, A. (2007) 'A joint motion & disparity motion estimation technique for 3D integral video compression using evolutionary strategy', *IEEE Transactions on Consumer Electronics*, **53**, 732–739.

[12] Davies, N. (1988) 'Three-dimensional imaging systems: A new development', *Applied Optics*, **27**, 4520–4528.

[13] Lippmann, G. (1908) 'Epreuves reversibles donnant du relief', *Journal of Physics*.

[14] Myungjin, C., Daneshpanah, M., Inkyu, M. and Javidi, B. (2011) 'Three-dimensional optical sensing and visualization using integral imaging', *Proceedings of the IEEE*, **99**, 556–575.

[15] Fusiello, A. Trucco, E. and Verri, A. (2000) 'A compact algorithm for rectification of stereo pairs', *Proceedings of Machine Vision Applications*, **18**, 16–22.

[16] Fecker, U., Barkowsky, M. and Kaup, A. (2008) 'Histogram-based pre-filtering for luminance and chrominance compensation of multi-view video', *IEEE Transactions on Circuits and Systems on Video Technology*, **18**, 1258–1267.

[17] Hirschmüller, H. (2008) 'Stereo processing by semiglobal matching and mutual information', *IEEE Transactions on Pattern Analysis and Machine Intelligence*, **30**, 328–341.

[18] Zilly, F. (2010) 'Specification of 3D production workflow and interfaces phase 1', Tech. Rep., D1.2.1, Multimedia Scalable 3D for Europe MUSCADE Integrated Project, June.

3

Compression

Compression is a substantial stage in the end-to-end 3DTV chain, where a huge load of raw video information needs to be carried over bandwidth-constrained networks. Compression had always played an important role in traditional video communication systems, by enabling the video information to be represented by tens of times lesser amount of bytes. Besides, the amount of visual information involved in 3D video communications is further multiplied compared to the 2D video, depending on the number of camera viewpoints that are delivered to the end users. The traditional video coding approaches applied to 2D video signals are also applicable to 3D video representation formats, despite their inability to remove the vast amount of redundancies in them. Thus, more sophisticated compression algorithms are needed for 3D video coding in order to extract the majority of inter-camera statistical correspondences and reduce the bit-rate of the resultant compressed representation.

This chapter is devoted to the widely deployed and modern 3D and multi-view video compression techniques. Video coding principles will be briefly provided that will be followed by the explanation of standards on 3D video coding. Finally, a detailed explanation on the most common forms of 3D video will be provided, i.e. stereoscopic, multi-view and multi-view with depth map.

3.1 Video Coding Principles

Transmission of raw video signal requires high bandwidths and excessive storage capacities which are unrealistic and costly. Therefore, video coding is required to compress the video while allowing the reversible conversion of data, requiring fewer bits and more efficient transmission. Video coding

3DTV: Processing and Transmission of 3D Video Signals, First Edition.
Anil Fernando, Stewart T. Worrall and Erhan Ekmekcioğlu.

targets exploiting the redundancies in the video sequence together with considering the Human Visual System (HVS) in order to achieve compression. The redundancies within a video sequence can be basically identified as spatial and temporal correlations. In the context of 3D video, the dimension of similarity is further extended to inter-camera statistical correspondences (texture and motion) as well as the similarities existing among different entities (such as camera view and depth maps).

Coding schemes exploiting the spatial redundancies are usually collected under the common name of intra-frame coding. Similarly, the schemes that make use of temporal redundancies existing among progressive frames of a video sequence are called inter-frame coding schemes. Temporal redundancies are reduced using the motion estimation process, which forms the basic block in inter-frame coding. This process is based on matching each block (quadratic pixel groups) with the candidate blocks in the reference frame and selecting the best matching block in the reference frame, where the reference frame is usually the adjacent frame in time. The best matching block can be chosen by considering several approaches including Mean Squared Error (MSE) and Mean of Absolute Error (MAE). The motion vector is the spatial displacement between the searched block and the best match according to the result of the search process.

Some spatial redundancies are exploited by predicting the values of each pixel or pixel group using previously coded information, and coding the remaining error (residual). This technique is called Differential Pulse Code Modulation (DPCM), which forms a very primitive form of video compression with relatively lower compression ratios compared to its competitors. Neighbouring pixels yield the best performance as they usually have the highest correlation with the pixel under consideration. Hence, the power of the produced error signal is low. The neighbouring pixels can be either from the same time-frame or from the adjacent frames in time. The prediction error (residual) is calculated and coded by subtracting the predicted frame from the original frame at the encoder and the received error signal is added to the prediction at the decoder. Note that this is a lossless coding scheme. However, it is also possible to introduce quantization and produce shorter code words. Accordingly, it is possible to make a DPCM encoder lossy in an adaptive way. Quantization in video coding is described as the process of determining the size of transmission unit (bits) for different parts of the signal, depending on the corresponding visual significance.

Among the other kinds of modern video coding schemes are segmentation (video object) based coding, wavelet coding (or in a generalized way, sub-band coding), model-based coding and transform coding. Segmentation-based encoders split the video frames into several regions, each of which has a different size and shape. The most commonly deployed approaches to extract different segments are based on the texture and motion information within the video scene. The texture colour (or luminance) information is exploited to generate fine video segments, whereas the motion data sought along the temporal axis helps to combine finer segments to yield a larger size and more compact segments. It is also common practice to pass the video frames through a non-linear filter to remove noises in the images that would result in detecting unnecessarily small and random segments. This should be non-linear and not applied through the objects' edges in the video. The shape and the texture information of each segment are encoded separately. This type of video coding is suitable for high compression applications that consider the human perception aspects. In other words, since each segment can be treated separately with different quantization step sizes, perceptually less significant video segments can be compromised in favour of more salient segments.

Sub-band coders decompose the video frames into its frequency components using a set of filter banks. The visual information carried in each sub-band frequency has a different level of saliency, considering the selective nature of HVS. Each sub-band component can be discarded from the reconstruction of the video frames that would result in a partial loss of quality. Thus, sub-band coders are inherently scalable, where it is desirable to first discard higher frequency sub-bands to which the human eye is less sensitive. Following the sub-band extraction, the coefficient for each sub-band is quantized. The lowest sub-band is usually predictive encoded, where the rest of the sub-bands are quantized relatively in a coarser way and run-length encoded.

Model-based (or model-aided) encoders are oriented around matching certain shapes in the video frames with previously computed model templates. Subsequently, the templates are deformed to match the exact shape existing in the video frames and the differences from the original model (i.e. the deformation factors) are encoded and sent. Both the encoder and the decoder pairs should know and use the same model templates. Such encoders are particularly successful in yielding very high compression ratios, provided that the video sequences comprise easily extractable foreground objects that closely match the pre-computed models.

Block-based transform video coding, which is one of the most widely used approaches, divides the video frame into hierarchical blocks of smaller size and maps the texture information of every block of each video frame into the transform domain prior to data reduction. This transform domain is called the Discrete Cosine Transform (DCT) domain and these blocks are referred to as macroblocks (MB). DCT is a very powerful transform, which condenses the discrete signals energy locally. In other words, the energy in the transform domain is unevenly distributed among bands. Most of the blocks' texture information tends to be accumulated in a few of the lowest sub-frequency components. Thus, in most cases, the energy of a video frame is concentrated in the low frequency band after DCT is applied to it. Those elements can be preserved by exploiting smaller quantization step sizes. On the other hand, insignificant transform elements (referred to as higher AC coefficients) can be discarded by harsher quantization. This process causes losses in the original data, as some high frequency coefficients in the transform domain are sometimes completely discarded.

Having outlined the basic highlights of the algorithms of modern video compression suits, the following section will give an insight into the well-known international video coding standards, in particular the MPEG-4 Part 10/H.264 Advanced Video Coding (AVC) standard of ISO and ITU. Note that in the section that describes the 3D video coding techniques, recent and emerging standardization efforts will be outlined.

3.2 Overview of Traditional Video Coding Standards

The first attempts at video coding systems date back to the 1960s when an analogue video phone system was designed. Organized groups of experts focused on video coding, thus the International Telecommunication Union (ITU-T) and the Joint Photographic Experts Group (JPEG) had made attempts at standardization in the late 1980s. Figure 3.1 illustrates the development of video coding standards starting from 1984.

H.261 (standardized in 1990), MPEG-1 Video (standardized in 1993), MPEG-2 Video (standardized in 1994), H.263 (standardized in 1997), MPEG-4 Visual (or Part 2) (standardised in 1998) and MPEG-4 Part 10/H.264 Advanced Video Coding (AVC) (standardised in 2004) are the existing coding standards used in multi-media applications, varying from low bit-rate communication to video telephony, from terrestrial broadcasting services to consumer video storage systems. Most of these coding standards rely on block-based transform coding and hybrid coding techniques making use of motion-compensated prediction.

The H.261 standard uses a combination of inter-frame Differential Pulse Code Modulation (DPCM) and Discrete Cosine Transform (DCT). It originally was intended to encode video sequences at 384 kbps, whereas later this was extended to operate at other bit-rates (multiples of 64 kbps). Coding technologies for video storage, such as CD-ROMs, were

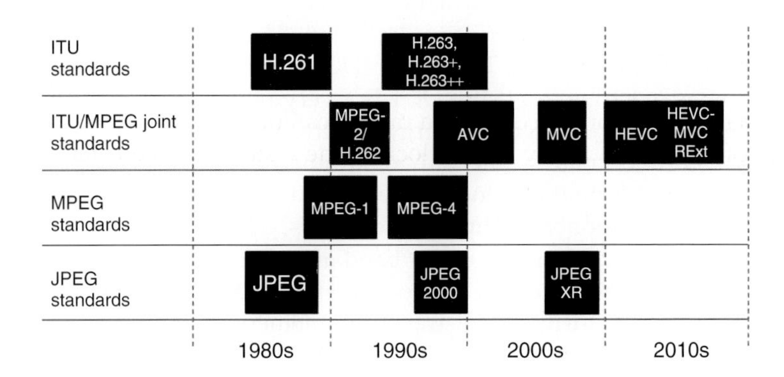

Figure 3.1 Development of video coding standards

investigated by the Motion Picture Experts Group (MPEG) in the early 1990s. The first generation of MPEG, MPEG-1 coding standard, uses H.261 as a starting point. MPEG-1 optimally operates in the range of 1.2–1.5 Mbps for non-interlaced video. Later, a new generation of standards emerged for coding interlaced video at higher bit-rates (in the range of 4–9 Mbps) called MPEG-2. MPEG-2 became so popular that it is used in many digital video applications such as terrestrial video broadcasting (DVB-T), satellite TV, cable TV and Digital Versatile Disc (DVD).

Later, MPEG-2 was adopted by ITU-T under the generic name of H.262. After progressive developments on MPEG-1 and MPEG-2, MPEG started working on a video-object-based coding standard and came up with MPEG-4. ITU-T, in parallel, carried out some work on a new video coding standard H.263 targeting coding at very low bit-rates. The evolutions of this standard were called H.263+ and H.263++, where the compression efficiency of this video coding standard was improved over the years. In 1997, two groups, ITU-T and ISO/IEC MPEG came together and formed a Joint Video Team (JVT) in order to create a single video coding standard: H.26L. Later, this codec was published jointly as Part 10 of MPEG-4 and ITU-T Recommendation H.264 [1] and called Advanced Video Coding (AVC). AVC is among the latest entries in the series of international coding standards [1]. AVC has achieved a significant improvement in rate-distortion efficiency relative to existing standards. It is noteworthy that AVC has almost double the compression performance of MPEG-2. In the following subsection, a brief overview of AVC is given.

3.2.1 Overview of MPEG-4 Part 10/H.264 AVC Standard

Described by its 'network-friendly' characteristic and high rate-distortion performance over heterogeneous networks, AVC is used in quite a wide range of video coding applications including broadcast over various media, multimedia data storage; conversational, multimedia streaming and multimedia messaging services on Ethernet, LAN, DSL, wireless and mobile networks, modems, etc. Figure 3.2 shows the high-level encoder block diagram of the AVC coding standard.

The frames of video are split into 16×16 pixel blocks called macroblocks. The macroblocks are encoded in a raster scan order from top left to bottom right of the frame. Every macroblock inside a single frame consist of three components: Luminance (Y), and two chrominance components Cr and Cb (representing colour). The chrominance samples are usually sub-sampled by a factor of two in both horizontal and vertical directions, since the human visual system is less sensitive to chrominance channels. Hence, a macroblock consists of 16×16 luminance and two 8×8 chrominance samples. The macroblocks are grouped as slices and there are five types of slices defined in AVC, which are Intra (I), Predictive (P), Bi-predictive (B), Switch-I (SI) and Switch-P (SP) slices.

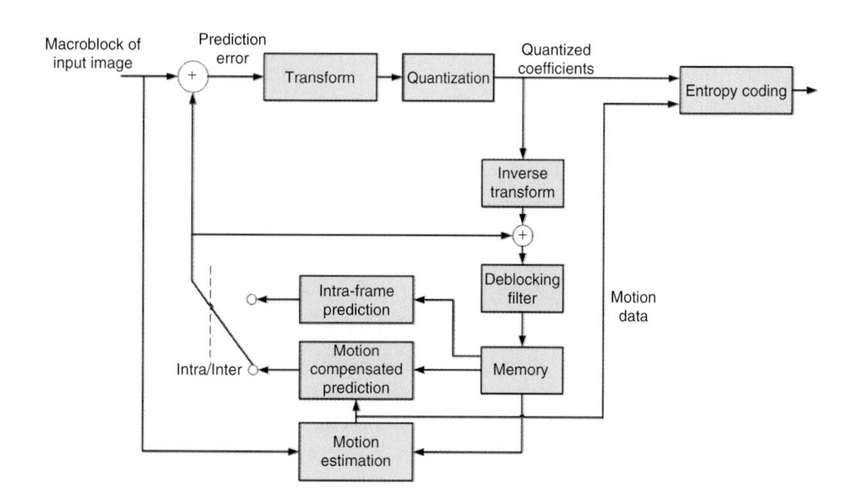

Figure 3.2 High-level encoder block diagram of AVC (3)

I-slices are encoded by intra-prediction, whereas P and B slices are encoded by inter-prediction. P slices can use references from a list of preceding reconstructed frames, whereas B slices can use references from two lists of reconstructed pictures: from preceding frames and from succeeding frames in time. The SI and SP slices are especially used for efficient bit-rate switching. More detail on SI and SP frames can be found in [2]. A single frame may contain a mixture of these slices.

Intra-coding is a scheme that does not exploit temporal redundancies existing among successive frames of a video sequence. The aim of intra-coding is to provide error resilience and create temporal random access points in a video sequence, more than to provide coding efficiency. Different from the previous coding standards, AVC applies a spatial estimation method for intra-coding. Spatial estimation is performed within the same frame using already encoded macroblocks. Intra-coding is especially efficient in compressing texture-wise smooth areas inside a frame.

Inter-coding aims to remove the temporal redundancies inside the video sequence via block motion estimation, as previously explained. There are two kinds of inter-frame coding in AVC: P (predictive) and B (bi-predictive) coding. In both techniques, the reconstructed samples from previously encoded frames are used. In previous standards, only the frame, which is the nearest frame to the currently encoded frame in time, could be used for P prediction. However, in AVC, multiple previously encoded frames can be used as the reference for P prediction. In B prediction, the encoded frames, which are encoded previously but succeeding the currently encoded frame in display time, can be used as references for prediction. However, this brings the cost of larger coding delays and increased memory usage, while yielding much higher compression performance.

As in previous standards, motion estimation is a non-normative method and not part of the standard, performed only during encoding. However, in AVC, different from the former standards, the size of the blocks used for motion estimation is not restricted to 16×16 and 8×8. Motion estimation can be carried out on 8×16 or 16×8 blocks as well. Besides, if 8×8 blocks are used, the motion estimation can be further carried out on 4×4 blocks. This increases the accuracy of the motion estimation. The accuracy of the estimated motion vectors is further increased by deploying quarter pixel motion estimation. However, this brings an additional computational load to both the encoder and the decoder. To encode the motion vectors, a spatial estimation is used, since the motion vectors are usually highly correlated in the space.

Examples of how the optimization is performed in conventional video coders, including AVC, can be found in [4]. Readers can refer to [5] to find more information on the motion-compensated prediction used in the AVC standard. After the estimation process, the residual macroblocks are formed by subtracting the predicted macroblock from the original macroblock pixel by pixel. Transform coding then is applied to the obtained residual macroblocks, in order to remove the spatial redundancies in the residual block.

In previous coding standards, a two-dimensional DCT of size 8×8 was applied. Integer transforms are performed in AVC in order to improve the computational performance of the codec platforms. The size of these transforms ranges from 8×8 to 4×4 mainly, and 2×2 (Hadamard Transform) for some cases (only for the DC coefficients – the lowest frequency band of 4×4 transformations) to better adapt the coding of residuals along the object boundaries. Different from DCT, since all the entries of integer transforms are integers ranging from -2 to 2, the transforms and inverse transforms can be applied easily by shifting, summation and subtraction operations. This fact makes the computational load of the encoding process decrease significantly. Furthermore, since this transform has an exact integer inverse transform, there is no possibility of a mismatch between the encoder and the decoder. Nevertheless, the energy compaction performance of the integer transforms is comparable to that of the DCT. All the coefficients of integer transforms are quantized by a scalar quantizer. There are 52 different quantization step sizes, which are denoted as the Quantization Parameter (QP). The quantization step size is doubled for every six increments in QP value.

The transformed and quantized coefficients of the residual macroblocks and the other syntax elements, i.e. the motion vectors, the indices of the reference frames, the prediction type of the macroblocks, are all entropy coded prior to transmission. AVC provides two options for entropy coding [3]: Context Adaptive Variable Length Coding (CAVLC) and Context Adaptive Binary Arithmetic Coding (CABAC). The key characteristic of both entropy coding schemes is that they dynamically update the codebook used to represent the video elements during encoding according to the changes in the content of the video.

CAVLC is computationally less intensive and simpler than CABAC, where CABAC generally saves more bits during encoding when compared to CAVLC with its 'non-integer length' code-word assignment capability. Both methods provide a significant increase in the compression with respect to the entropy coding methods used in former video coding standards. Huffman coding is one example of the variable length entropy coding techniques used in former coding standards.

3.2.2 High Efficiency Video Coding (HEVC)

Recent standardization efforts have yielded the new High Efficiency Video Coding (HEVC) (also commonly referred to as H.265, or MPEG-H Part 2), which is expected to be more efficient than its predecessor AVC. Similar to the AVC standard, HEVC is also a project of the Joint Collaborative Team on Video Coding (JCT-VC) of ISO/IEC-MPEG and ITU-VCEG. HEVC aims to provide significantly improved compression efficiency compared to the AVC High Profile. The intention is to reduce bit-rate requirements by up to 50% with comparable image quality [6]. The committee draft standard was published in February 2012, while the final international draft standard was released in January 2013. This section provides a brief overview of some of the new features in HEVC, and further details concerning HEVC can be found in [7].

One of the prominent features of the HEVC model is the use of variable block sizes in a tree structure, ranging from 8 × 8 to 64 × 64, unlike AVC that uses fixed size macroblocks of size 16 × 16. The largest blocks are called Coding Tree Blocks (CTBs), whose size may be changed from sequence to sequence. CTBs may be specified as being 16 × 16, 32 × 32, or 64 × 64. CTBs may be further segmented into Coding Units (CUs). CUs' sizes range from 8 × 8 to 64 × 64. An example of how a CTB may be split into different CU sizes is shown in Figure 3.3.

Coding Units can be segmented into Prediction Units (PUs), where each segment of a PU may be given a different Motion Vector. The Inter PU types specified in HEVC are shown in Figure 3.4. The bottom half of Figure 3.4 shows the partitions that may be obtained using the Asymmetric Motion Partitioning (AMP) option, which divides a block into two unequal parts (a three quarters – one quarter split). Two new tools are employed to make motion representation more efficient: Advanced Motion Vector Prediction (AMVP) and Merge mode. AMVP allows more efficiency prediction and coding of motion vectors, by allowing a predicted motion vector to be formed from a list of candidate prediction motion vectors. Candidates are derived from temporally co-located PUs in other frames, and from spatially adjacent PUs.

Merge mode allows motion vectors to be predicted by using an index value to select from particular motion vector predictors. This allows the decoder to predict and derive the motion vectors for Merge mode coded PUs. Together these tools make it possible to describe more detailed motion

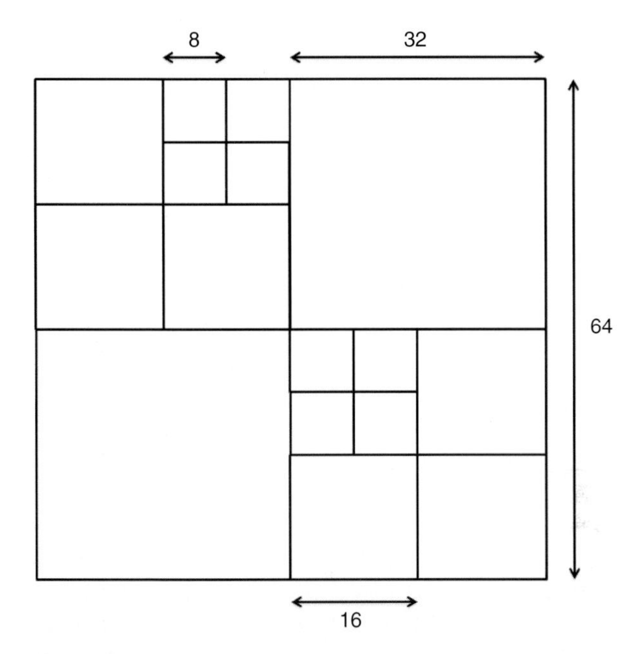

Figure 3.3 Example of how a CTB may be split to contain different size Coding Units in HEVC, where each square represents a different Coding Unit

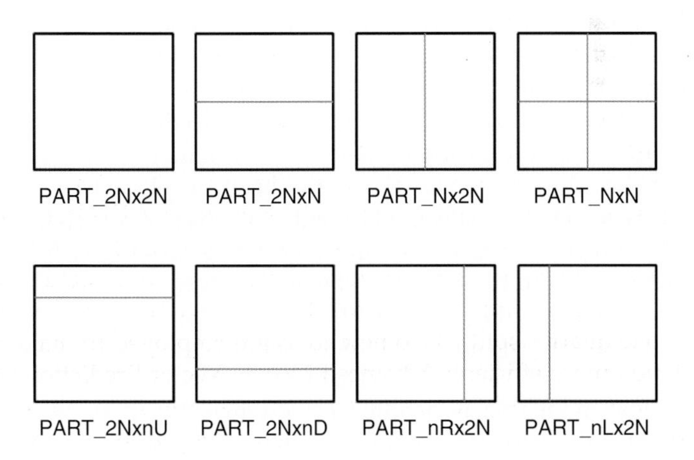

Figure 3.4 Prediction Unit types specified in HEVC for Inter CU coding

information, without the large amount of motion vector bits that would be required in AVC.

Intra CUs may only use $2N \times 2N$ and $N \times N$ partitions. For Intra CUs, PUs describe the Intra prediction mode used in a particular area, rather than a Motion Vector. HEVC defines 35 different intra-frame prediction modes for improved intra-frame coding performance [8]. Combined with the flexible tree-based partitioning, this gives a significant compression efficiency gain (around 25%) compared to AVC Intra coding. The strong performance of the Intra coding tools has led to the definition of a 'Still Image' profile, allowing HEVC to be used for image compression.

HEVC allows different size Transform Units (TUs), which means that different size integer transform sizes are used, from 4×4 up to 32×32. Each CU contains a transform tree, allowing different size transforms to be used within a CU. Note that this also means that a transform must be equal to or smaller than the size of the CU. Due to their larger size, and larger multipliers, HEVCs transforms are significantly more complex for some implementations than AVC.

Transform skip is another interesting new feature in HEVC. Transform skip involves all of the normal processing for Intra- and Inter-coded CUs, but allows the transform process to be bypassed. This feature may only be used with 4×4 transforms. It seems to have advantages for text-based content, and allows ringing around text to be significantly reduced. For coding of screen-based content (e.g. PowerPoint presentations), the gain in quality is very significant (>11%).

Compared to AVC, significant changes have been made to the loop filtering. Similar to AVC, HEVC features an in-loop deblocking filter, which has been refined and improved for inclusion in the new standard. HEVC also features an additional loop filter: Sample Adaptive Offset (SAO) [9]. SAO applies simple pixel offsets to the reconstructed picture, to provide a reference frame that more closely matches the original picture. An efficient syntax for encoding the pixel offsets has been developed as part of the standard. An encoder can specify different SAO offsets for each CTU.

HEVC uses a context-adaptive binary arithmetic coding (CABAC) algorithm that is similar to the CABAC scheme deployed in the AVC standard. On the other hand, unlike in AVC, CABAC is the only entropy encoder scheme that can be used in HEVC. AVC allows context-adaptive variable length coding (CAVLC) and CABAC to be two options at the entropy coding stage. Context modelling is handled more efficiently in HEVC's CABAC scheme that results in significant improvements in the throughput. Different from the binary arithmetic coding method applied in AVC, the bypass-mode within CABAC has also been improved in HEVC to further increase the throughput [10].

The motion compensation filters of HEVC also differ from the filters employed in AVC, where either half-sample or quarter-sample precision filters with eight taps can be selected. AVC uses only a half-sample precision

6-tap filter. In addition to the increase in taps, HEVC always uses 14-bit accuracy in motion compensation, irrespective of the bit-depth of the video source material.

3.3 3D Video Coding

L-R stereoscopic video, the simplest form of 3D video, requires more storage capacity and higher bandwidth for transmission compared to 2D video. Therefore, 3D video coding is crucial to make the immersive video applications available for the mass consumer market in the near future. The coding approaches for 3D video may be diverse depending on the representation of 3D video. 3D video coding approaches aim to exploit inter-view statistical dependencies in addition to the conventional 2D video coding approach, which removes the redundancies in the temporal and spatial dimensions. The prediction of views utilizing the neighbouring views and the images from the same image sequence are shown in Figure 3.5. The efficiencies of the prediction methods shown vary depending on the frame rate, the inter-camera baseline distances and the complexity of the content (e.g. spatial and temporal characteristics).

3.3.1 Stereoscopic Video Coding

Many stereoscopic video coding algorithms have been proposed and tested till now. A common approach has been to encode one of the viewpoints (either left-eye or right-eye view) conventionally using conventional video coding tools (e.g. AVC), and the other view using disparity compensated prediction. This is a fundamental way to exploit the 3D scene geometry in improving the rate-distortion efficiency of the stereoscopic video codec. Several disparity-compensated prediction-based stereoscopic video coding works exist in the literature, such as in [11] and [12]. It is important to

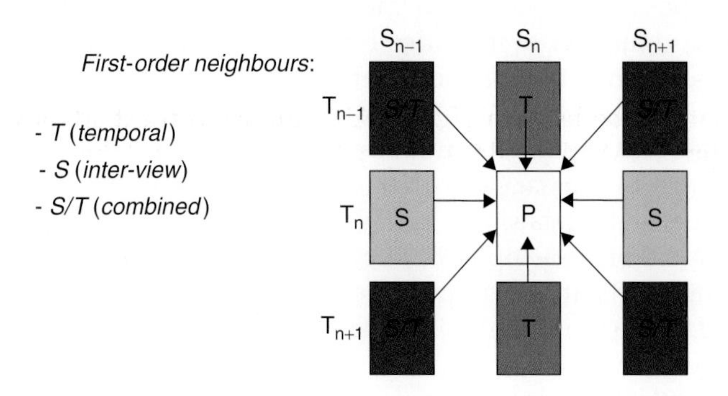

Figure 3.5 Statistical dependencies in a three-dimensional video

note that the disparity-compensated prediction scheme is not applicable to video-plus-depth type 3D video, since the texture characteristics of colour and depth videos are different.

Asymmetric [13] and mixed-resolution coding techniques are also available to left-eye- and right-eye-based stereoscopic video coding, based on the suppression theory of human visual perception. The suppression theory of human visual system states that the overall perception of the stereo-vision is primarily affected by the highest resolution of the video frame sequences arriving at either eye [14]. Accordingly, a significant amount of bits can be saved by encoding one of the viewpoints either with reduced spatio-temporal resolution or with lower fidelity. Such coding approaches are further applicable to video-plus-depth type 3D video and accordingly, the depth map transmission overhead can be maintained at around 20% of that of the colour texture video with pleasant output 3D video quality.

MPEG-4 Multiple Auxiliary Component (MAC) [15] allows image sequences to be coded with a Video Object Plane (VOP) on a pixel-by-pixel basis, which contains data related to video objects such as disparity, depth and additional texture. However, this approach needs to send the greyscale/alpha shape with any other auxiliary component (e.g. disparity) and as a result, the coding efficiency is negatively affected.

One of the major challenges in designing compression suits for stereoscopic video is to carefully analyse the effect of colour perception and depth perception on the overall stereoscopic video quality, to correctly decide on the trade-off between colour texture and depth map video coding rates. Usually, coding depth map information with less accuracy does not affect the stereoscopic reconstruction quality as much as coding the colour texture video with less accuracy. Furthermore, due to the characteristics of the depth image sequence (e.g. large smooth image areas), they can be efficiently compressed with the existing coding algorithms. Cues of the human visual system need to be taken into account carefully while assigning different quantization step sizes, in stereoscopic or multi-view video coding, in order not to introduce eye-fatigue problems after viewing for a long time.

It is also possible to signal the method(s) used to pack the frames of a stereoscopic video pair in the video bit-stream, utilizing supplemental user data (e.g. Supplemental Enhancement Information (SEI) messages defined in the AVC standard). The AVC standard has already defined such an SEI message that signals the frame packing method. The Frame Packing Arrangement SEI message tells the decoder that the left and right view frames are packed into a single high-resolution video frame (e.g. HD frame compatible) either in a top-to-bottom, side-by-side, checkerboard, or any other arrangement. It is also possible to signal the temporal interleaving scheme used for left and right view frames that are not packed into a single HD frame (e.g. Full Stereo).

Packing both views into a single video frame makes it possible to use existing AVC conformant hardware equipment, decoders and set-top boxes to decode 3D video streams immediately without having to upgrade them to decode other 3D video specific standards such as AVC Stereoscopic High Profile. It is possible to exploit the frame compatible 3D video services as the base layer of a more extensive 3D video service, where the additional enhancement layer would exploit spatial prediction from the frame compatible base layer at the lower spatial resolution to produce Full HD stereoscopic output. Interested readers can consult [16], where a detailed description of frame compatible stereoscopic representation and coding methods, as well as detailed information on various SEI messages defined in the AVC standard to assist stereoscopic video applications are given.

As previously described in Chapter 2, to realize immersive applications that comprise a more realistic replication of the captured 3D scene, more than stereoscopic representation is required. This usually is possible by capturing the scene with a series of conventional cameras (more than two), on a pre-constructed and well-calibrated platform (i.e. a rig). The following section outlines the properties of the multi-view coding extension of the AVC standard, which is specialized to take into consideration the inter-view statistical correlations while removing most of the redundancies existing between different camera views.

3.3.2 Multi-View Video Coding

If two viewpoints in a stereoscopic video are separately encoded and transmitted, as much as twice the bandwidth that would be required for a 2D video has to be allocated for the transmission. Likewise, if there are N total viewpoints recorded, in total, N times the total bit-rate necessary to send a single view would be required, where N may be substantially large according to the demand of the application. It is inevitable that more sophisticated compression schemes for multi-view sequences will need to be exploited.

ISO/IEC MPEG has recognized the importance of multi-view video coding and established an ad-hoc group (AHG) on 3D audio and visual (3DAV) in December 2001 [17]. Four main exploration experiments have been conducted by the 3DAV group. The experiments were conducted between 2002 and 2004. These experiments included:

1. Exploration experiment on omnidirectional video.
2. Exploration experiment on FTV.
3. Exploration experiment on coding of stereoscopic video using the multiple auxiliary component of MPEG-4 (mentioned previously).
4. Exploration experiment on depth and disparity coding for 3DTV and intermediate view synthesis [18].

After the Call for Comments issued in October 2003, several companies called for the need for a standard enabling the FTV and 3DTV systems. Subsequently, MPEG called interested parties to produce evidences on

Multi-View Coding (MVC) technologies in October 2004. Some evidences were recognized in January 2005 and a Call for Proposals on MVC was officially issued in July 2005. Then, the responses to the Call were evaluated in January 2006. Finally, MVC was standardized in July 2008.

Several requirements have been set for modern multi-view coding systems. Some requirements are identified as the compression efficiency, scalability in viewpoint direction, spatial, temporal and SNR scalability. In addition, a multi-view coding system must be backward compatible. Coding and decoding with low delay is another requirement of multi-view coding to enable real-time collaborative and interactive 3D applications. Since the visual data load under concern is high, the issue of view, temporal and spatial random access is also critical and any multi-view coding suit should have low delay random access characteristics. Besides, the decoder resource management issue needs to be taken into account when designing a multi-view coding algorithm. Parallel processing of different views or segments of the multi-view video is also important to facilitate efficient encoder and decoder implementations. These issues are essential in particular to deploy multi-view applications in mobile devices.

The methods applied within the context of exploiting the inter-view redundancies can be separated into two classes: inter-view redundancy removal via reference frame-based techniques and inter-view redundancy removal via disparity-based techniques.

The techniques based on reference frames selected from different views adapt the existing motion estimation and compensation algorithm to remove the inter-view redundancies. Basically, the disparity among different views is treated as motion in the temporal direction and the same methods are applied to model disparity fields. One example of such a multi-view video coding method, introduced in [19] and called the Hierarchical B-Frame Prediction uses the same hierarchical decomposition structure, applied in the temporal domain, in the view domain.

Figure 3.6 shows the prediction structure used in this specific multi-view coding method [19]. The horizontally directed arrows in Figure 3.6 represent referencing in the time domain, whereas the vertically directed arrows in red represent referencing in the view domain. In this way, a frame being encoded may have references from both the same view or from neighbouring views. In particular, the pictures belonging to the highest temporal and view level (the light blue frames in B-predicted view) are the most efficiently coded frames inside the multi-view sequence, since they are predicted from both temporal references and inter-view references. However, more memory is required for reference frames and more dependent frames must be decoded before the frame in the highest temporal and view level can be decoded. One drawback of such an encoding structure is that there is no distinction between the motion in time and the disparity among the frames of different views. The motion fields and disparity fields have different characteristics. In general, motion in time is bounded within a certain search field and changes

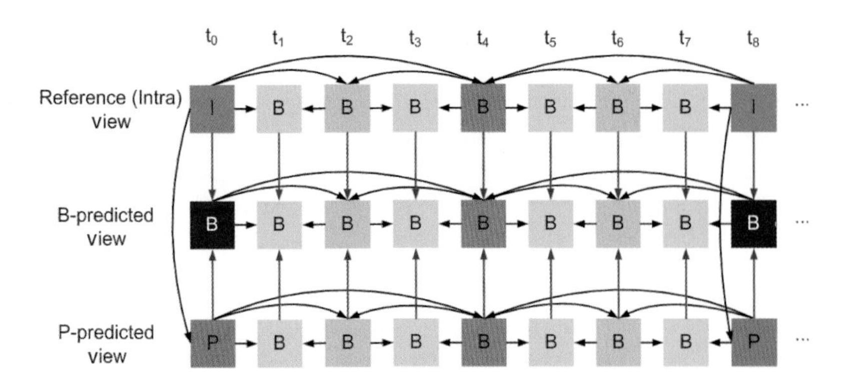

Figure 3.6 Generic prediction structure for the MVC method

dynamically within the same frame. However, disparity among views is more dependent on the distance between the cameras and can have much greater values than the motion vectors. Furthermore, there generally exists a global disparity between the views, which does not change in time, unless the multiple cameras are moving.

Experiments have been performed with different multi-view sequences to analyse the efficiency of interview prediction structures in [20]. It has been shown that temporal prediction is the most efficient prediction mode for all analysed sequences on average. The relationship between the temporal and the inter-view prediction strongly depends on the scene complexity and the temporal/spatial density. However, in the results shown in [20], it is seen that inter-view prediction attains a quality improvement of 0.5–2 dB for most of the sequences over coding each view separately with the hierarchical B-frame prediction. However, this gain changes from one prediction structure to another. Furthermore, it is seen that when there is too much disparity between neighbouring camera views, the encoder cannot fully exploit the redundancies.

The basic prediction structures are shown in Figure 3.7. In Figure 3.7 (a), the simulcast coding structure is shown. This shows that no inter-view prediction is used. In Figure 3.7 (b), inter-view prediction is used only for key frames. In Figure 3.7 (c), inter-view prediction is used for non-key frames as well as the key frames. The prediction structure denoted in Figure 3.7 (c) is found to be the best performer (1.7 dB gain on average). However, it achieves a small gain over the structure denoted in Figure 3.7 (b) for a much increased complexity. It is interesting to note that the prediction structure with bi-predictive coding applied to key frames performs worse in general when compared to the scheme with no bi-predictive coding applied to key frames. This is due to the QP cascading scheme used in hierarchical B prediction. With QP cascading, there is less fidelity for lower hierarchy levels. For instance, B Frames are encoded with a higher QP than P frames and I frames.

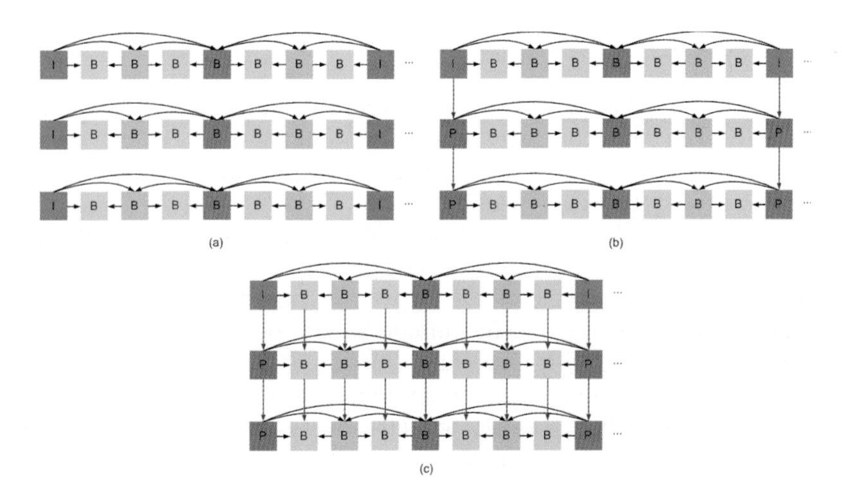

Figure 3.7 Various inter-view prediction structures

In terms of coding complexity, it is noted that the prediction structures that include inter-view compression for non-key frames do slightly better than the prediction structures that do not include inter-view prediction for non-key frames, but at around three times more complexity. The optimum prediction structure is dependent on the scene complexity and the camera arrangement. Nevertheless, it is impossible to avoid a linear increase in the transmission bandwidth when the total number of viewpoints is increased, no matter which prediction structure explained above is used.

Another approach for multi-view coding takes into account the geometric constraints and tries to improve prediction performance by exploiting the scene geometry. These approaches are classified as disparity-based approaches. Almost all of these techniques rely on generating intermediate representations of frames (sometimes just part of a frame) to be used as prediction sources in addition to inter-view, intra-frame and inter-frame prediction sources. Generating these representations can be carried out in two ways: using the scene geometry information either implicitly or explicitly [21].

In coding scenarios where the scene geometry is explicitly used, the encoder makes use of the depth information of the scene to synthesize prediction images. In [22], the authors exploit view synthesis prediction in addition to inter-frame and inter-view predictions. Generally, in view synthesis processes, either the depth information is available for every pixel location of different views or the depth information is generated using disparity matching techniques. In [22], novel views are rendered via 3D image warping at the camera positions where the view to be encoded is located.

In coding scenarios, where the scene geometry is implicitly used, i.e. no depth information is applied for synthesis, the positional correspondences between neighbouring cameras are used [21]. View interpolation techniques can be classified in this class. In [23], the authors introduced the 'Group of Group Of Pictures' (GoGOP) concept for inter-view prediction. Figure 3.8 shows the overview of such a GoGOP structure. The authors allowed only the decoded pictures in the GoGOP to be referred to during prediction [23]. In [24], the authors have proposed including the synthesized pictures, which are interpolated using the neighbouring views frames at the same time instant, in the GoGOP structure. Figure 3.9 shows the reference frame diagram for the proposed method in [24]. In Figure 3.9, $D(c_a, t_i)$ represents an decoded picture of view c_a at time t_i and $M_n(c_a, t_i)$ represents an interpolated picture, where $n(0 \leq n < N)$ represents the number of interpolated pictures and N is the total number. $M_n(c_a, t_i)$ are produced from the decoded picture of the left view at the same instant, t_i, and the decoded picture of the right view at t_i. According to this method, the distance between the encoded view and the left neighbour view need not necessarily be the same as the distance between the encoded view and the right neighbour view. The advantage of this method

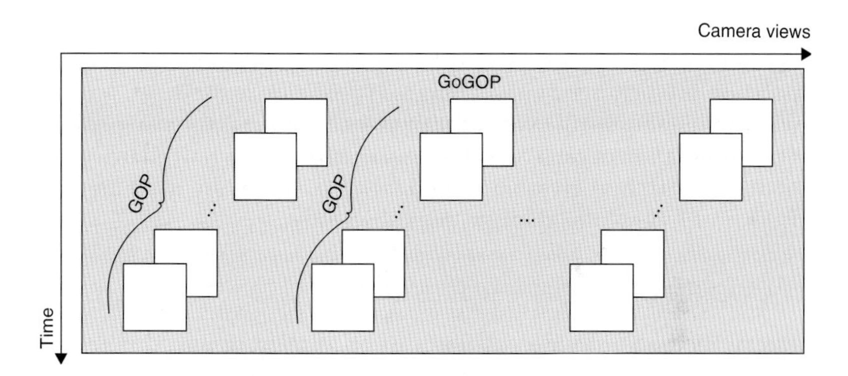

Figure 3.8 GoGOP structure presented in (23)

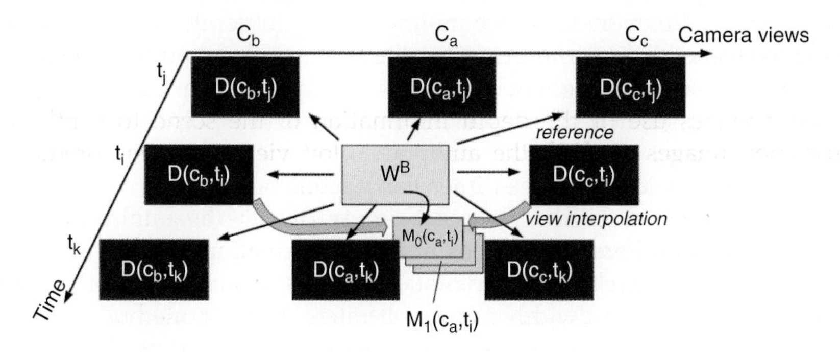

Figure 3.9 Reference picture selection for the method presented in (24)
Source: IEEE Xplore

is that, neither disparity maps nor depth maps need to be sent to the decoder side. The camera parameters do not need to be known. Only an interpolation parameter needs to be encoded and sent with a considerable lower overhead. This parameter defines the relative positions of the neighbouring cameras used in the view interpolation process to the camera whose view is encoded. The disadvantage of this method on the other hand is that the approach of view interpolation is not suitable for high quality view rendering at on the decoder side for display. Furthermore, the view interpolation quality is strictly dependent on the distance between the encoded view and the neighbouring view. As the distance under concern increases, the estimation quality of the interpolated view decreases.

In [25], a multi-view coding approach is used for multi-video plus associated depth maps kind of data format. The main concern in this work is to use an MVC method to be able to render high quality display videos in real time. In [25], two cameras inside a multi-view sequence are selected as base views, and both their colour texture videos and depth maps are encoded separately. The remaining views and their depth maps are spatially predicted from the reconstructed versions of the base views. Finally, the created residuals are encoded and sent to the decoder side. In this approach, no means of view interpolation prediction or view synthesis prediction is used.

Having discussed the 3D video formats that comprise the per-pixel dense depth representation of each camera view, i.e. the multi-view plus depth map format, the next section gives an insight into the commonly used methods to compress multi-view plus depth videos efficiently.

3.3.3 Coding of Multi-View Plus Depth

Traditional methods of video coding can be applied to encode depth map sequences with high efficiency [26]. Other techniques have been proposed and implemented, owing to its distinct characteristics. A 3D motion estimation approach has been proposed in [27]. In this method, the best motion vector is found based on an exhaustive search in 2D image dimensions and the Z (depth) direction. Since the motion activity in depth maps is not only restricted to two dimensions, but three, the motion estimation/compensation process of the underlying video codec can be improved to take into consideration the motion of a macroblock in the Z direction (pointing in or out of the display plane).

Figure 3.10 depicts the three-dimensional motion estimation concept. This method gives good results at high bit-rates due to the minimization of residual energy. However, it does not perform well at low bit-rates due to the increase in the size of the motion vectors, which constitute a large proportion of the bandwidth at low bit-rates. Another method that was described in [28] suggests sharing the motion vectors between the encoded colour texture video and the corresponding encoded depth map video. However, the correlation of the motion vector sets of both videos depends

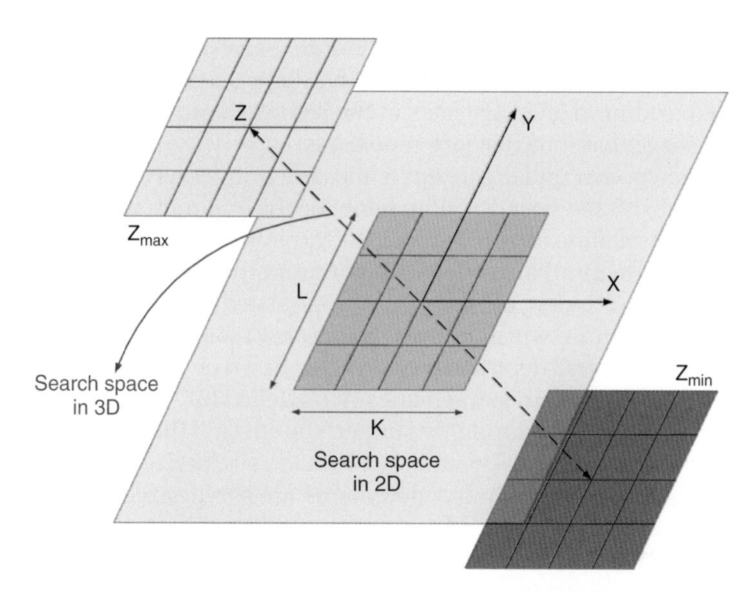

Figure 3.10 Three-dimensional motion estimation concept for coding depth maps, presented in (27)

heavily on the content of the video. Hence, if the correlation is low, the energy of the encoded residuals is high that results in the consumption of more bits.

Reduced resolution coding techniques are also commonly used for video-plus-depth and multi-view depth map coding systems [29–31]. The effects of data loss due to downsampling, especially in various depth regions, have not been thoroughly studied in these works. In [32], the authors presented a novel concept for depth map coding based on platelets, assuming that the depth map frames are piecewise smooth images. They compared their results against AVC intra-coding, which outperformed the rate-distortion performance of the codec based on platelets.

The wide viewing angle coverage and capability of rendering intermediate viewpoints make multi-view plus depth format stronger than other previously preferred 3D formats for 3D free-viewpoint video applications. Multi-view coding tools are equally applicable to coding the multi-view depth map and this has been done in this way in some previous research work [33–35].

A different aspect that comes onto the scene in multi-view plus depth coding is the effect of the compression-related distortions in depth maps on free-viewpoint rendering. Accordingly, several approaches address the issue of allocating total transmission bandwidth among colour texture viewpoints and depth map viewpoints, to achieve improved arbitrary viewpoint synthesis quality [31, 36]. The traditional way of allocating bit-rate among colour texture viewpoints and depth map viewpoints with a fixed percentage (usually 20% of the bit-rate used for colour texture video encoding is allocated to depth maps) yields consistent results with stereoscopic viewpoint generation, due to the restricted baseline distance view generation. However, in the context of multi-view free-viewpoint video, the synthesis process becomes more sensitive to coding-related artefacts in the reconstructed depth map frames, due to the larger baseline distance rendering range. The visual quality of the arbitrary viewpoints under concern is measured in terms of PSNR, in most of the work discussed here. PSNR may not be the optimum way of assessing the perceived quality of synthesized viewpoints, as the synthesis-related geometrical (structural) distortions are not properly assessed by PSNR.

Some other multi-view plus depth coding-related research works exist in the modern literature, which address the issue of joint encoding of colour-texture viewpoints and depth map viewpoints based on novel view synthesis distortion models that are mathematically modelled [37, 38].

Authors in [39] have thoroughly studied the effect of different depth compression levels (in terms of quantization) on the synthesized virtual camera viewpoint quality. The arbitrary viewpoints are synthesized at evenly distributed places between two real camera viewpoints. The quality assessment is based on PSNR and the reference videos are generated using original (uncompressed) colour texture videos and depth maps. Figure 3.11 depicts the assessment process for depth map coding performance. The authors compared the compression of their platelet-based depth map encoder with a multi-view coding (MVC) scheme for encoding the multi-view depth maps, as well as with the intra-coding mode of AVC standard (neither inter-view, nor inter-frame prediction allowed).

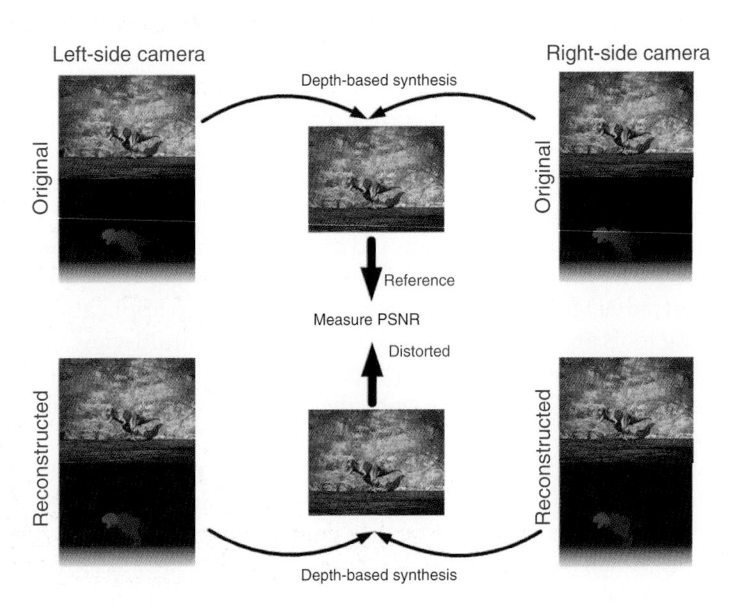

Figure 3.11 Evaluation of view synthesis rendering quality with compressed depth video (as presented in [39])

According to the test results, it has been found that MVC that is based on AVC beats the platelet-based multi-view depth codec, as well as the AVC intra-coder, with a large margin in terms of rate-distortion performance. It is, however, the case, if the distortion of depth map reconstruction is taken into account during the in-loop rate distortion optimized mode selection process. On the other hand, when the reconstruction quality of arbitrarily synthesized virtual viewpoints is taken into account, the performance of the platelet-based coder beats that of the AVC intra-coder and comes close to the performance of the AVC-based MVC at various depth map coding rates. It is necessary to preserve the edge information in depth maps, in order to facilitate better rendering of views, where edges in depth maps correspond to sharp depth discontinuities. Taking into account subjective quality assessment, the authors come to the conclusion that the platelet-based multi-view depth map coder preserves the sharp depth discontinuities more than MVC does, and hence introduces less coding-related structural artefacts in the synthesized viewpoints.

In [40], the authors compare two different mode decision techniques in depth map encoding using the AVC standard, where the distortion measure for depth maps was selected as the distortion in the resultant synthesized views using the corresponding compressed depth maps. Both of the compared depth-block coding mode selection approaches relied on the mentioned depth/disparity quantization error metric, as shown in Figure 3.12, which takes into consideration the amount of error introduced in the synthesized view. Subsequently it has been discovered that taking the distortion measure of a compressed depth map as the resulting structural error in the

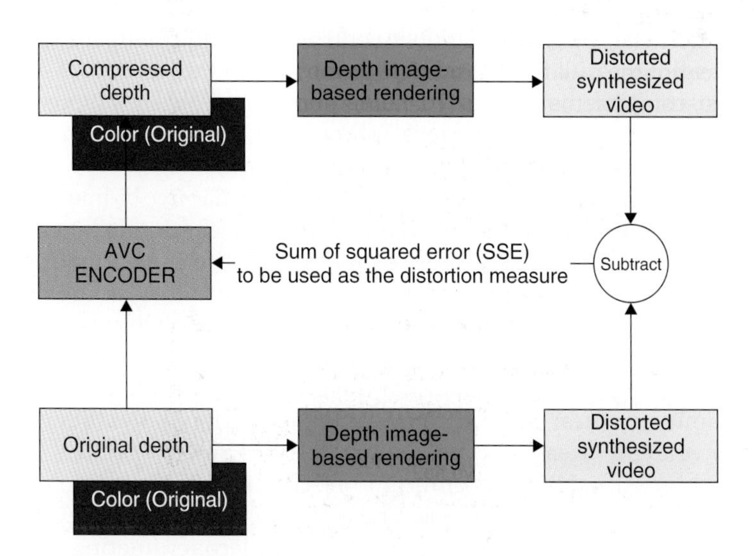

Figure 3.12 Depth distortion measure as the SSE between the original and distorted synthesized view (as described in (40))

synthesized camera views leads to up to 2 dB quality improvement. This was verified using a number of test 3D video sequences, operating the encoder at several low, medium and high source bit-rates.

Usually, it is a convenient approach to process depth maps/multi-view depth map sequences, so that they become easily compressible and at the same time, become effective in avoiding synthesis artefacts. Deployment of such processing algorithms can be avoided by using robust depth map estimation algorithms, despite their relatively high complexity and iterative characteristics that seriously hinder real-time processing. Such processing approaches do usually take into account the influence of different depth regions (depth layers in 3D) on view synthesis distortions. It is well known that depth layers that lie far away from the capturing camera, unlike the depth layers closer to the capturing camera, have negligible effect in creating visual holes (occluded regions) as well as structural ambiguities in object borders. This fact entrains the necessity of designing locally adaptive processing approaches for depth maps, as applied in [41].

In addition, sharp depth discontinuities, i.e. foreground–background object transitions, require special attention in order to avoid the incorrect hole-filling process after view synthesis. A post-decoding processing framework had been implemented for depth maps to improve the foreground–background transition fidelity that is a result of compression artefacts, in [42]. In [43], the authors proposed a method to process compressed depth maps using an adaptive bi-linear filter to improve the perceptual quality of the view synthesis. As is known, bi-lateral filters are edge-preserving low-pass filters that comprise a similarity and a proximity kernel (modelled as two-dimensional Gaussian filters in this work) for weighting the samples. Adaptation is controlled by controlling the similarity component of the bi-lateral filter, in such a way that the centre and standard deviation components of the similarity kernel have been adapted to the particular depth pixel's luminance following a depth histogram analysis and the extraction of depth value peaks in the depth image.

The flow diagram of the depicted post-decoding adaptive bi-linear filtering is shown in Figure 3.13. After the identification of every dominant peak with its surrounding valley, the nearest peak for each processed depth pixel is determined. Subsequently, the adaptive bilateral filter weights are assigned for the processed depth pixel. Such an adaptation scheme is motivated by the fact that depth maps are grey-scale image sequences that comprise large smooth regions inside, and that the grey-values inside a depth image are accumulated near several dominant levels. Thus, after filtering, the majority of depth areas that are characterized as smooth regions are successfully blurred, while the significant object transitions are sharpened, on the contrary. The depth compression results with the standard AVC encoder reported in [43] have shown that when the post-decoder filtering is applied on the decompressed depth map images, the stereoscopic rendered video sequences quality can be improved by up to 1.9 dB on average.

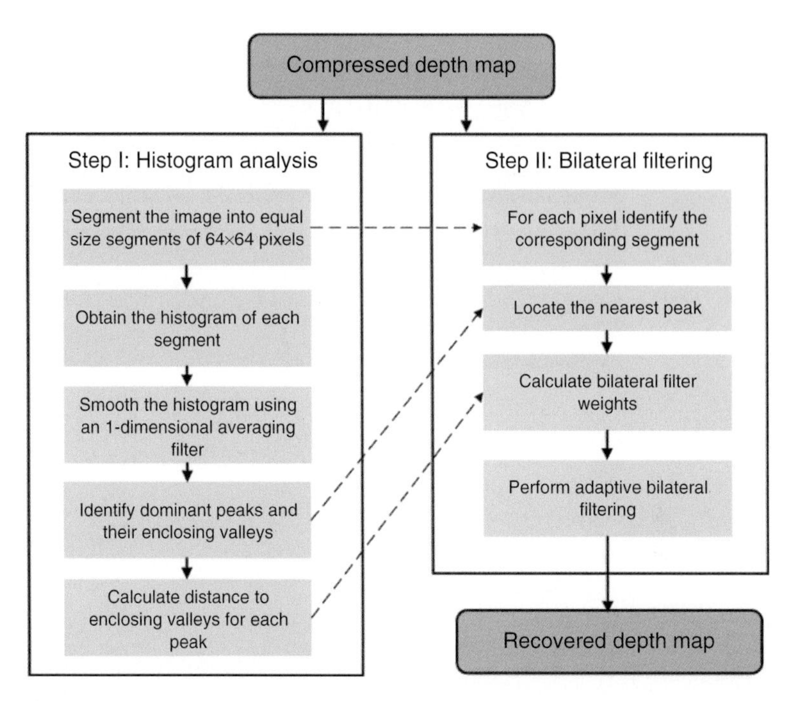

Figure 3.13 Block diagram of the adaptive depth post-processing framework (taken from (43))

In [44], the authors applied a similar idea in the depth pre-encoding stage, in order to remove unnecessary spatial variations in the depth map sequences that would have a not significant influence on the view synthesis quality. However, by removing the non-required high frequency elements in the depth map sequences, some useful bit-rate is saved. On the other hand, significant depth transition areas are preserved by properly adjusting the filter coefficients. Different from the work presented in [43], the work reported in [44] uses a colour-video aided tri-lateral Gaussian filter instead of a bi-lateral filter. This involves an additional filter kernel, which takes into account the similarity of colour and corresponding depth video edges. At every depth map pixel to be processed, a window of $2w \times 2w$ is formed centred at the corresponding depth pixel. Subsequently, in these kernels of $2w \times 2w$ (denoted by Ω), the filtered depth value is computed as

$$D_p = \sum_{q \in \Omega} coeff_{pq} \cdot I_q / \sum_{q \in \Omega} coeff_{pq} \tag{3.1}$$

where, q denotes a pixel within the kernel and p is the centre pixel to be processed. The *coeff* is a multiplication of three different factors, namely the closeness in pixel, similarity in depth value and the similarity in colour

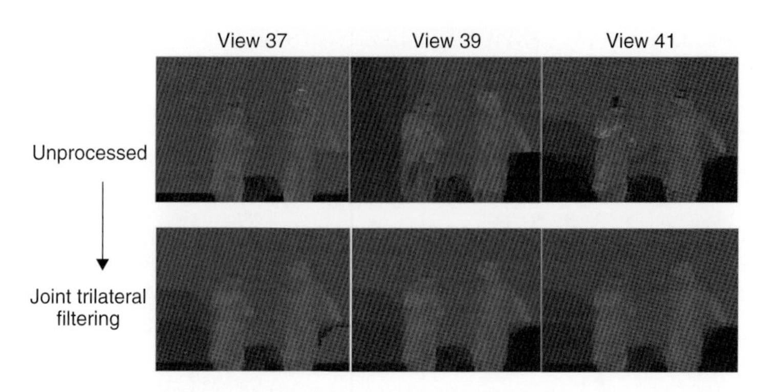

Figure 3.14 Multi-view depth maps before and after tri-lateral filtering (Reproduced with the permission of Masayuki Tanimoto)

texture value, respectively:

$$coeff_{pq} = c(p,q) \cdot s_{depth}(p,q) \cdot s_{colour}(p,q) \tag{3.2}$$

The filters related to these three factors are considered as Gaussian filters centred at point p. Accordingly, these individual factors are denoted as:

$$\begin{aligned} c(p,q) &= \exp\left(-\frac{1}{2}(p-q)^2/\sigma_c^2\right) \\ s_{depth}(p,q) &= \exp\left(-\frac{1}{2}\left(d_p - d_q\right)^2/\sigma_{s_{depth}}^2\right) \\ s_{colour}(p,q) &= \exp\left(-\frac{1}{2}\left(I_p - I_q\right)^2/\sigma_{s_{colour}}^2\right) \end{aligned} \tag{3.3}$$

where, d represents the depth values of pixel points p and q, and I represents the corresponding colour texture luminance values. Figure 3.14 shows an example snapshot from a test multi-view depth map sequence (three views in a row), before and after the application of the colour-aided tri-lateral filter. It is seen that the majority of textural gradients in the original depth maps are smoothened, while the depth edge transitions are preserved. Reported coding results have suggested that applying the processing filter on the extracted multi-view depth map video sequences can reduce the bit-rate for encoding the depth maps effectively by up to 52%.

3.4 Recent Trends in 3D Video Coding

In March 2011, MPEG released a new call for proposals for encoding generic 3D video signals, which can be considered a follow-up to the multi-view coding standard (MVC) that had already attracted many researchers and multimedia technology companies worldwide. The MVC standard proved to be a success and superior to the native AVC standard as regards stereoscopic or multi-view compression. From the coding efficiency point of view, as the number of possible representations of a certain scene increases, i.e. different viewing angles, the amount of bit-rate budget to represent the multi-view video increases and hence, more sophisticated methods should be employed to compress the content. Future 3DTV systems will require multi-view coding schemes, which are able to compress the 3D multi-view video content to be transmitted over band-limited networks under heavy load, while providing interoperability and scalability at the same time.

On the other hand, the current MVC approach is conceptually unable to keep the 3D video transmission load at manageable rates, while the number of camera views to shoot a particular scene keeps increasing for higher 3D resolution (targeting multi-view displays). Hence, it is not possible to commercially realize 3DTV systems based on multi-view video services that consist of extremely many cameras and use the current compression technology. Table 3.1 outlines the estimate source coding bit-rate for various 3D video formats (at HD resolution and 25 fps, progressive coding) with three state-of-the-art video coding standards. Note that HEVC is planned be standardized in early 2013 following the development work. It can be seen that as long as the number of camera views involved in the 3D video format increases, the bit-rate of all encoders tends to increase linearly.

Based on the rate constraint, a newer 3D video framework has been introduced by the standardization body MPEG that puts a limit on the input number of cameras [45]. The 3D video framework suggests ideally coding two viewpoints or at most three viewpoints and sufficiently producing as many viewpoints as needed at the client (decoder) side. This suggestion is also in line with the limitations applicable in the widely deployed modern production platforms, which can accommodate a restricted number of cameras in a compact, calibrated and easily manageable manner. The main purpose is to set an upper limit on the required bit-rate to transmit the corresponding 3D video format. In other words, for a given resolution, as

Table 3.1 Indicative bit-rates necessary to transmit various MVV formats using state-of-the-art video coding standards

Video Codec	Resolution@ Frame-rate (Pixels@ Frames/second)	2D (One view) (Mbit/s)	Stereoscopic (Mbit/s)	Four Views (Mbit/s)	Eight Views Mbit/s
AVC	$1920 \times 1080p@25$	10-20	20-40	40-80	80-160
MVC	$1920 \times 1080p@25$	10-20	14-28	28-56	56-112
HEVC	$1920 \times 1080p@25$	5-10	10-20	20-40	40-80

well as for any 3D display content requirement, the transmission of 3D video services should not incur substantial rates overhead, but at the same time should produce high enough quality output 3D video.

For widely manufactured and used stereoscopic 3D displays, variable stereoscopic baseline and adjustable depth perception are the desirable features that could be realized by streaming the appropriate multi-view plus depth content. Similarly, for multi-view auto-stereoscopic displays, which can display more than two camera views simultaneously (real and synthesized views), the main desired features are wide viewing angle and a large number of dense synthesized views. Thus, the main technical challenges in such a framework lie in determining the optimal baseline distance between the encoded viewpoints, so that there is enough correlation between the two viewpoints to encode them in a joint and efficient way. At the same time, sufficiently many arbitrary viewpoints should be rendered in between them. In other words, 3D video framework aims to achieve an optimum trade-off between the bit-rate necessary to transmit the 3D video representation and the synthesis performance of multiple intermediate viewpoints.

Following publication of the 3D video framework, the standardization bodies have agreed a path forward that will deliver a large variety of ways to compress 3D video. The workplan is outlined in [46]. The standards being worked on can be categorized as using AVC or HEVC as the basis for compression.

3.4.1 3D Video with AVC-Based Coding Technology

AVC-based approaches will build on the existing AVC and MVC standards to allow coding of video using depth maps. This will extend both AVC and MVC to handle depth maps. Section 3.3.3 has already discussed how MVC can be improved by using depth maps. Both AVC and MVC extensions will support coding of depth at a different spatial resolution to the colour texture. Coding depth maps at lower resolutions has been found to be a good way to improve compression efficiency. The depth maps are not directly viewed by end-users. As long as the rendered colour content is high resolution, it is possible to code depth maps at a lower resolution.

The AVC approach allows predictions to be made between depth and colour texture in a flexible manner. The MVC approach codes depth and texture independently. It has been reported that the AVC approach can actually deliver better performance than the MVC-based approach, with gains reported to be around 23%.

The MVC plus depth standard (known as MVC+D) will be completed in January 2013, while the AVC-based approach (known as 3D-AVC) is expected to be completed in November 2013.

3.4.2 3D Video with HEVC-Based Coding Technology

Two different HEVC-based approaches are being advanced. The first, known as MV-HEVC, will extend the HEVC standard to allow coding of multiple views, without depth maps. The standard will exploit correlation between views in a similar way to the MVC standard. It is expected that a final version of MV-HEVC will be completed in January 2014.

The second approach, known as 3D-HEVC, will try to use depth maps in a similar fashion to the 3D-AVC standard. In other words, the specification will exploit correlations between depth and colour video, and will allow coding of depth with a lower resolution compared to the colour video. It is expected that 3D-HEVC will provide around 25% better compression efficiency than MV-HEVC.

References

[1] Wiegand, T., Sullivan, G. J., Bjontegaard, G. and Luthra, A. (2003) 'Overview of the H.264/AVC video coding standard', *IEEE Transactions on Circuits and Systems for Video Technology*, **13**, 560–576.

[2] Karczewics, M. and Kurceren, R. (2003) 'The SP- and SI-frames design for H.264/AVC', *IEEE Transactions on Circuits and Systems for Video Technology*, **13**, 637–644.

[3] Ostermann, J., Bormans, J., List, P., Marpe, D., Narroschke, M., Pereira, F., Stockhammer, T. and Wedi, T. (2004) 'Video coding with H.264/AVC: Tools, performance, and complexity', *IEEE Circuits and Systems Magazine*, **1**, 7–28.

[4] Sullivan, G. and Wiegand, T. (1998) 'Rate-distortion optimization for video compression', *IEEE Signal Processing Magazine*, November.

[5] Ghanbari, M. (2003) *Standard Codecs: Image Compression to Advanced Video Coding*, London: Institution of Engineering and Technology.

[6] ISO/IEC JTC1/SC29/WG11 Coding of Moving Pictures and Audio, 'Vision, applications and requirements for high efficiency video coding (HEVC)', Tech. Rep. N11872, Video and Requirements Subgroups and JCT-VC, 2011.

[7] Sullivan, G.J., Ohm, J-R., Han, W-J. and Wiegand, T. (2012) 'Overview of the High Efficiency Video Coding (HEVC) standard', *IEEE Transactions on Circuits and Systems for Video Technology*, **22**, 1649–1668.

[8] Lainema, J., Bossen, F., Han, W-J., Min, J. and Ugur, K. (2012) 'Intra coding of the HEVC standard', *IEEE Transactions on Circuits and Systems for Video Technology*, **22**, 1792–1801.

[9] Fu, C.-M., Alshina, E., Alshin, A., Huang, Y-W., Chen, C-Y., Tsai, C-Y., Hsu, C.-W., Lei, S-M., Park, J.-H. and Han, W.-J. (2012) 'Sample adaptive offset in the HEVC standard', *IEEE Transactions on Circuits and Systems for Video Technology*, **22**, 1755–1764.

[10] Sze, V. and Budagavi, M. (2012) 'High throughput CABAC entropy coding in HEVC', *IEEE Transactions on Circuits and Systems for Video Technology*, **22**, 1778–1791.

[11] Balamuralii, B., Eran, E. and Helmut, B. (2005) 'An extended H.264 CODEC for stereoscopic video coding', *Proceedings of SPIE – The International Society for Optical Engineering*, pp. 116–126.

[12] Moellenho, M.S. and Maier, M.W. (1998) 'DCT transform coding of stereo images for multimedia applications', *IEEE Transactions on Industrial Electronics*, **45**, 38–43.

[13] Aksay, A., Bilen, C., Kurutepe, E., Ozcelebi, T., Akar, G.B., Civanlar, M.R. and Tekalp, A.M. (2006) 'Temporal and spatial scaling for stereoscopic video compression', in *Proceedings of IEEE European Signal Processing Conf. EUSIPCO 2006*, Florence, Italy, September.

[14] Stelmach, L.B. Tam, W.J. Meegan, D. and Vincent, A. (2000) 'Stereo image quality: Effects of mixed spatiotemporal resolution', *IEEE Transactions on Circuits Systems for Video Technology*, **10**, 188–193.

[15] ISO/IEC 14496-2 (2001) 'Generic coding of audio-visual objects part 2: Visual', Tech. Rep., Doc. N4350.

[16] Vetro, A., Wiegand, T. and Sullivan, G. (2011) 'Overview of the stereo and multi-view video coding extensions of the H.264/MPEG-4 AVC standard', *Proceedings of the IEEE*, **99**, 626–642.

[17] ISO/IEC JTC1/SC29/WG11 (2001) 'List of ad-hoc groups established at the 58th meeting in Pattaya', Tech. Rep., N371.

[18] Smolic, A. and McCutchen, D. (2004) '3DAV exploration of video-based rendering technology in MPEG', *IEEE Transactions on Circuits and Systems for Video Technology*, **14**, 348–356.

[19] Schwarz, H., Hinz, T., Smolic, A., Oelbaum, T., Wiegand, T., Mueller, K. and Merkle, P. (2006) 'Multi-view video coding based on H.264/MPEG4-AVC using hierarchical B pictures', in *Proceedings of Picture Coding Symposium*, China.

[20] Merkle, P., Smolic, A., Muller, K., and Wiegand, T. (2007) 'Efficient prediction structures for multiview video coding', *IEEE Transactions on Circuits and Systems for Video Technology*, **17**, 1461–1473.

[21] Shum, H., Kang, S., and Chan, S. (2003) 'Survey of image based representations and compression techniques', *IEEE Transactions on Circuits and Systems for Video Technology*, **13**, 1020–1037.

[22] Martinian, E., Behrens, A., Xin, J., and Vetro, A. (2006) 'View synthesis for multiview video compression', in *Proceedings of Picture Coding Symposium*, China.

[23] Kimata, H., Kitahara, M., Kamikura, K. and Yashima, Y. (2004) 'Multi-view video coding using reference picture selection for freeviewpoint video communication', in *Proceedings of Picture Coding* Symposium, Lisbon, Portugal, December.

[24] Yamamoto, K. (2007) 'SIMVC: Multi-view video coding using view interpolation and color correction', *IEEE Transactions on Circuits and Systems for Video Technology*, **17** (11), 1436–1449.

[25] Zitnick, C.L. (2004) 'High-quality video view interpolation using a layered representation', *ACM Siggraph and ACM Transactions on Graphics*, **23**, (3), 600–608.

[26] Fehn, C. (2004) 'Depth-image-based rendering (DIBR), compression and transmission for a new approach on 3DTV', in *Proceedings of SPIE Conference on Stereoscopic Displays and Virtual Reality Systems XI, 5291*, CA, USA, pp. 93–104, January.

[27] Kamolrat, B., Fernando, W. and Mrak, M. (2008) '3D motion estimation for depth information compression in 3D-TV applications', *IET Electronic Letters*, **44**, 1244–1245.

[28] Grewatsch, S. and Miller, E. (2004) 'Sharing of motion vectors in 3D video coding', in *IEEE International Conference on Image Processing*, Singapore, pp. 3271–3274.

[29] Ekmekcioğlu, E., Worrall, S.T. and Kondoz, A.M. (2008) 'Low-delay random view access in multi-view coding using a bit-rate adaptive downsampling approach', in *Proceedings of IEEE International Conference on Multimedia and Expo*, pp. 745–748, June.

[30] Karim, H.A., Worrall, S. and Kondoz, A.M. (2008) 'Reduced resolution depth compression for scalable 3D video coding', in *Proceedings of Visual Information Engineering*, Workshop on Scalable Coded Media Beyond Compression, Xian, China, July.

[31] Ekmekcioğlu, E., Worrall, S. and Kondoz, A.M. (2008) 'Bit-rate adaptive down-sampling for the coding of multi-view video with depth information', in *Proceedings of 3DTV Conference: The True Vision: Capture, Transmission and Display of 3D Video*, Istanbul, Turkey.

[32] Morvan, Y., Farin, D. and de With, P.H.N. (2007) 'Depth-image compression based on an R-D optimized quadtree decomposition for the transmission of multiview images', in *IEEE International Conference on Image Processing*, San Antonio, TX, September.

[33] Merkle, P., Smolic, A., Muller, K. and Wiegand, T. (2007) 'Multi-view video plus depth representation and coding', in *Proceedings of IEEE International Conference on Image Processing 2007*, October.

[34] Klimaszewski, K., Wegner, K. and Dománski, M. (2009) 'Distortions of synthesized views caused by compression of views and depth maps', in *Proceedings of 3DTV-Conference 2009, The True Vision: Capture, Transmission and Display of 3D Video*, Potsdam, Germany, May.

[35] Tikanmäki, A., Gotchev, A., Smolic, A. and Müller, K. (2008) 'Quality assessment of 3D video in rate allocation experiments', in *Proceedings of IEEE International Symposium on Consumer Electronics (ISCE'08)*, Algarve, Portugal, April.

[36] Morvan, Y., Farin, D. and de With, P.H.N. (2007) 'Joint depth/texture bit-allocation for multi-view video compression', in *Proceedings of Picture Coding Symposium*, Lisbon, Portugal, November.

[37] Liu, Y. (2009) 'Compression-induced rendering distortion analysis for texture/depth rate allocation in 3D video compression', in *Proceedings of IEEE Data Compression Conference*, pp. 352–361.

[38] Liu, Y. (2009) 'Joint video/depth rate allocation for 3D video coding based on view synthesis distortion model', *Proceedings of Signal Processing: Image Communications*, **24**, 666–681.

[39] Merkle, P., Morvan, Y., Smolic, A., Farin, D., Müeller, K., de With, P. and Wiegand, T. (2009) 'The effects of multiview depth video compression on multiview rendering', *Signal Processing: Image Communication*, **24**, 73–88.

[40] Silva, D.D. and Fernando, W. (2009) 'Intra mode selection for depth map coding to minimize rendering distortions in 3D video', *IEEE Transactions on Consumer Electronics*, **55**, 2385–2393.

[41] Ekmekcioğlu, E., Velisavljevic, V. and Worrall, S. (2010) 'Content adaptive enhancement of multi-view depth maps for free viewpoint video', *IEEE Journal of Selected Topics in Signal Processing*, **5**, 352–361.

[42] Silva, D.D., Fernando, W., Kodikaraarachchi, H., Worrall, S. and Kondoz, A. (2011) 'Improved depth map filtering for 3D-TV systems', in *2011 IEEE International Conference on Consumer Electronics (ICCE)*, pp. 645–646, January.

[43] Silva D. D., Fernando, W., Kodikaraarachchi, H., Worrall, S. and Kondoz, A. (2011) 'Adaptive sharpening of depth maps for 3D-TV', *IET Electronics Letters*, **46**, 1546–1548.

[44] Ekmekcioğlu, E., Worrall, S., Velisavljevic, V., Silva, D.D. and Kondoz, A. (2011) 'Multi-view depth pre-processing using joint filtering for improved coding performance', Tech. Rep., ISO MPEG Doc m20070, Geneva, March.

[45] V.R. Group (2008) 'Vision on 3D video', Tech. Rep., ISO/IEC JTC1/SC29/WG11 N10357, February.

[46] Ohm, J-R., Rusanovskyy, D., Vetro, A. and Müller, K. (2012) 'Work plan in 3D standards development', Tech. Rep. JCT3V-B1006, Joint Collaborative Team on 3D Video Coding Extension Development, October.

4

Transmission

4.1 Challenges of 3D Video Transmission

Multimedia transmissions over communication networks are prone to transmission errors in the form of packet losses, bit errors, or burst errors. Therefore, the reconstructed video frames at the receiving end may differ from the original transmitted video content. Furthermore, these errors on reconstructed video propagate from one frame to another due to the prediction mechanism used in conventional video compression algorithms. This in turn degrades the perceived quality of the reconstructed video at the receiver. Similarly, the emerging immersive video communication applications will be affected by channel error conditions. Transmission of 3D video is more difficult compared to 2D video due to the involvement of the depth information. In any 3D video transmission, both 2D video (colour or texture) and depth are needed to be transmitted. This means if one sequence is corrupted, the final quality will be disrupted. Therefore, when designing 3D video transmission system, efficient resource allocation between 2D video (colour or texture) and depth needs to be carefully considered since the overall bit allocation is a limited resource.

4.2 Error Resilience and Concealment Techniques

With the growth in the range of multimedia services being used for everyday activities such as teleconferencing, mobile television and peer-to-peer video sharing, the reception of video with high quality is of prime importance to users, as well as to service providers. New video coding standardizations, such as H.264, scalable video coding, wavelet coding and distributed video coding, were introduced to accomplish these requirements. Some of these

3DTV: Processing and Transmission of 3D Video Signals, First Edition.
Anil Fernando, Stewart T. Worrall and Erhan Ekmekcioğlu.

techniques concentrate on video transmission between the servers and clients in heterogeneous networks, such as scalable extension of H.264/Advanced Video Coding (AVC), while others such as distributed video coding are more applicable to video uplink applications.

However, providing video communication over wireless or wired networks creates many challenges due to fluctuations in the channel characteristics. In internet packet network scenarios the whole packet can be lost during the transmission due to congestion, faulty network connections and signal degradation. In addition, bit errors caused by noisy channels and multipath propagation play a crucial role in mobile wireless transmission environments. These errors create artefacts in the reconstructed video frames that propagate in both the spatial and temporal domain due to the hierarchical prediction scheme employed in video compression stages. Therefore, it is essential to take the necessary precautions to mitigate these adverse effects. This in turns opens up opportunities for error resilient techniques.

Error resilience for video transmission has become a crucial area for research in the field of multimedia communication in the last decade. The methods used for this purpose vary from retransmission on request, embedding extra information in the coded stream to conceal the errors with already available data. Moreover, in error resilience approaches, measures were taken to reduce the amount of dependency between adjacent frames to reduce the error propagations.

However, the utilization of error-resilient tools in transmission networks is restricted by the channel bandwidth. The complexity of the codec and the loss of compression efficiency, also restrict the use of error-resilient techniques in some application scenarios. Therefore, it is essential to maintain a flexible balance between the error-resilient tools and the quality of the reconstructed video sequence. In addition, researchers have a couple of options when it comes to introducing new error resilience tools. That is either to improve the existing methods or to introduce new techniques that are compatible with existing video coding standards or invent novel coding architectures which have a good resistance to channel errors.

4.2.1 Background

Due to the dynamic nature and the unpredictability of the transmission channels, the bit stream transmitted through communication networks always undergoes either burst errors or bit errors. However, at the decoding end, the packets with bit errors will be dropped, resulting in corrupted pictures and deterioration in the quality of the output bit stream. Therefore, research on error recovery, concealment and error propagation prevention methods is an important aspect of multimedia communications.

In current practice there are several methods to mitigate these adverse effects, such as forward error correction (FEC), feedback-related algorithms, error concealment and error resilience mechanisms. However, the introduction of these error control and recovery mechanisms heavily degrades the coding efficiency. In general, there are four types of data losses associated with bit errors which occur in wireless video transmission systems:

1. *Least significant data loss* – These are the type of errors which occur at less important portions of the encoded segments. For example, errors in the texture data of the video will not propagate in the temporal domain, so they will only deteriorate the quality of the picture which they belong to. Also, a bit error in texture data will not propagate spatially. Hence, this is a localized error which causes the least significant data loss.
2. *Prediction data loss* – If bit errors take place in motion vectors, they will result in prediction errors. There is a higher tendency for these prediction errors to propagate in the temporal domain.
3. *Data loss due to synchronization loss* – Errors which occur in variable length code words can sometimes cause the decoder to lose synchronization. In such a case, the decoder is unable to detect the specific error location and it discards even the error-free portion of the received data until it reaches the next resynchronization marker. Therefore, this kind of bit error will turn into a burst error at the decoder.
4. *Loss of entire frame* – Bit errors occurring in the header data create the worst damage compared to the above-mentioned three types. When an error occurs at header data, the decoder completely loses the track of the encoder and in turn discards a whole video frame. Hence, this is the worst type of error that can happen in video transmission.

4.2.2 Error Resilience Tools

In a broad sense there are four types of error resilience tools associated with the video encoding and decoding. Namely:

1. Localization techniques.
2. Data partitioning techniques.
3. Redundant coding techniques.
4. Concealment-driven coding techniques.

It should be noted that these tools are not mutually exclusive. The introduction of these techniques will increase the bit-rates of the encoded sequence and the complexity of the codec, while decreasing the coding efficiency.

4.2.2.1 Localization

In this method the spatial and temporal dependency between the frames or slices has been removed or reduced to stop further propagation of the errors.

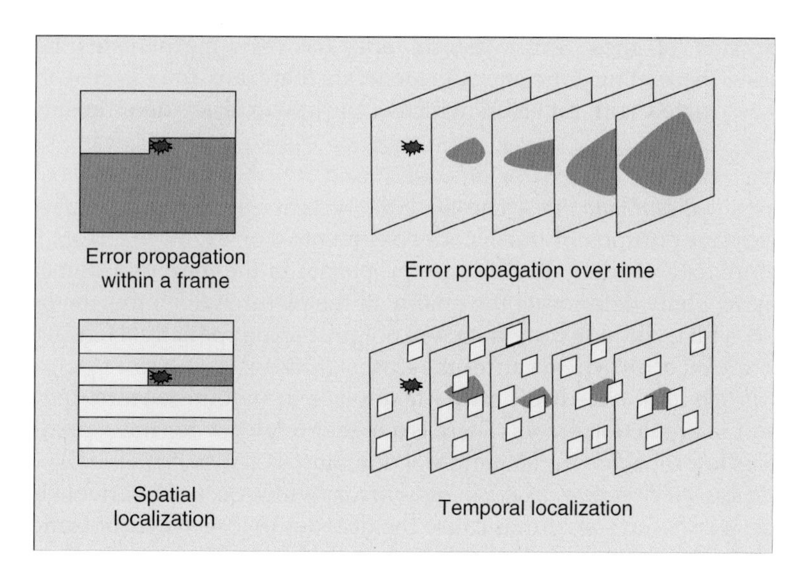

Figure 4.1 Illustration of spatial (left) and temporal (right) localization

Figure 4.1 depicts the impact of the spatial error propagation and the effect of spatial error recovery tools. Some of the well-known *spatial localization* techniques are:

- *Slice structure* – In this method, the whole picture is partitioned into a number of slices. Then these slices are transmitted as independent NALUs. Therefore, the error in one slice will not propagate in the spatial domain, as shown in Figure 4.1.
- *Resynchronization markers* – Spatial dependency between the adjacent MB can be limited by adding resynchronization markers at MB boundaries in predefined intervals. This in turn improves the synchronization between the segments in the bit stream. However, every one of these segments carries separate header information to help the restarting of the decoding. For instance, bit pattern 0x00000001 is used as the synchronization marker in the H264/AVC compatible data stream. The introduction of additional data to the bit stream results in a decrease in the coding and bandwidth efficiency. This is one of the key inherent disadvantages of this method.

In *temporal localization*, measures are taken to prevent error propagation in the time domain. Temporal error propagation happens as a result of errors in the prediction data. Hence, as an error prevention and recovery method the picture buffer is refreshed at regular intervals and the references back to erroneous pictures are minimized. Here are some examples of temporal localization:

- *Intra coded MB and Instantaneous Decoder Refresh (IDR) MB* – Error localization is achieved by intra coding frames in the sequence at selected intervals. Although intra coded MBs are not a specific error resilience tool, they are widely employed in encoding to prevent error propagation. In extreme cases where the channel is highly error prone, all the MBs of the video stream will be intra coded. Intra coding comes with the disadvantage of an increase in the output bit-rate. Therefore, it is wise to select these two methods at the encoder either according to a cyclic pattern or randomly or adaptively depending on the characteristic of the transmission channel.
- *Reference picture selection* – In this method, the encoder is customized to select only the frames decoded without any errors as future references to prevent error propagation. For effective functioning of the above method, it is essential to have a feed-back channel between the encoder and the decoder. The reference picture selection method, however, is supported only in the H.263 and MPEG-4 codecs.

4.2.2.2 Data Partitioning

In data partitioning, the encoded data of a frame is grouped or fragmented according to its importance. Because of this feature, in the case of an error in low priority data sections, the decoder can drop the corresponding packets and continue with decoding, provided that the group with the highest priority arrives without any errors. For example, in the H.264/AVC specification the encoded MB data is split into three groups: A, B, and C, depending on the relative importance of the encoded data. Group A consists of Header information, motion vectors and quantization parameters. Group B consists of intra coefficients, while the group C contains inter coefficients.

After the partitioning, data transmission can be prioritized according to the relative importance of the segments, such in the presence of a single channel the most crucial data can be transmitted with high amount of channel coding. Further, when multiple channels are available for transmission, the packets with the highest priority can be transmitted through the most reliable channel, facilitating unequal error protection to the encoded data.

4.2.2.3 Redundant Coding

In the redundant coding method additional data will be encapsulated with the bit stream to facilitate robust decoding at the decoder. This can

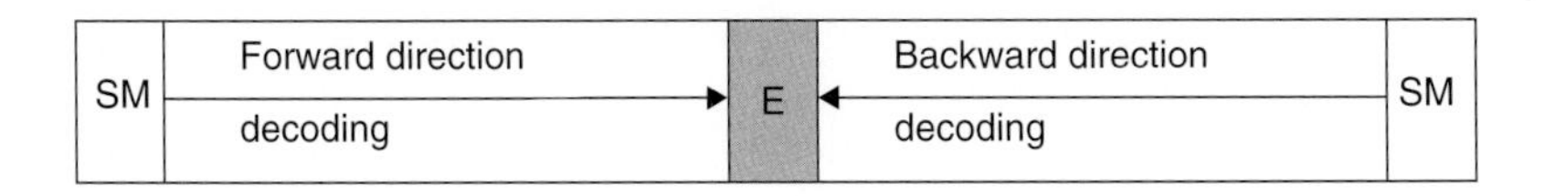

Figure 4.2 Decoding process of RVLC

be achieved explicitly by embedding duplicates of the primary slices as redundant slices, or implicitly by means of either Reversible Variable Length Coding (RVLC) or Multiple Descriptive Coding (MDC).

- *Redundant coding* – In this tool, the replicas of the primary coded pictures (PCP) will be encoded and transmitted along with the primary data stream. At the decoder the redundant pictures are decoded if and only if the primary pictures are lost, otherwise the redundant stream is discarded. Therefore, the redundant coding is extremely disadvantageous in error-free environments. Hence, measures have been taken to minimize the amount of channel bandwidth allocated for the redundant frames.
- *Reversible Variable Length Coding* – RVLC provides error resilience by means of both forward and backward decoding techniques. In other words, if an error (E) takes place between two adjacent synchronization markers (SM) in a bit stream, in general the whole data between the two markers will be discarded, regardless of the accuracy of the rest of the data. However, this can be overcome by the use of RVLC, which has the capacity to carry out decoding in both forward and backward directions as shown in Figure 4.2. The code words of the Variable Length Coder (VLC) must be symmetric, so that it can be decoded either in the forward direction or in the backward direction. This in turn limits the flexibility of VLC and decreases the efficiency of the codec.
- *Multiple Descriptive Coding (MDC)* – In this technique the bit stream is divided into mainly two streams which can be decoded independently to give a reasonable quality output video. The two streams are transmitted either in two different channels or multiplexed into one channel. At the decoder a reasonable quality image sequence can be regenerated if one of the streams is received without any errors. Moreover, if both streams arrive at the receiving end without any errors, then the decoder is able to generate an output image sequence with superior quality. Furthermore, the redundancy in the transmission data can be minimized by analysing the correlation presence between the two streams, at the encoder.

4.2.2.4 Concealment Driven Tools

As the name implies, in concealment driven tools the errors are concealed by means of the additional information received about the general behaviour of the lost data. These tools can be either encoder driven or decoder driven. Interpolation is a common method which has been used for the lost data prediction.

Examples of error concealment methods used in AVC are:

1. *Frame copy* – In this algorithm the error is concealed by copying the first picture in the reference list 0 of the corresponding sample. It can be performed on both base layer and enhancement layer levels.

2. *Temporal direct motion vector generation* – In this method all the motion vectors and reference indices for the missing picture are generated in a similar pattern to the temporal direct mode algorithm.
3. *Motion and residual upsampling (BISkip)* – The residual data and motion information is generated by upsampling the corresponding information of the base layer while the motion compensation is performed at the enhancement layer. Therefore, it is obvious that this algorithm can be used only to conceal errors at the enhancement layer.
4. *Reconstructed Base Layer upsampling (RU)* – As the name implies, this is an error concealment algorithm for the enhancement layers. In this method, the reconstructed base layer image is upsampled using 6-tap filters to recreate the lost frame.
5. *Flexible Macroblock Ordering (FMO)* – In FMO, the MB data will be organized into different slice groups. Then these groups will be re-arranged according to a predefined pattern which is known to both the encoder and the decoder, and are transmitted in separate packets. At the receiving end, the decoder re-arranges the scattered data before decoding. The missing portions, i.e. the erroneous segments, are predicted from the correctly decoded segments of the video sequence. By using this method, the system can minimize losses and perform predictions more accurately than the frame concealment option. In the H.264/AVC standard there are seven predefined patterns and the pattern "type1" gives the best performance of all these patterns.

4.2.3 Forward Error Correction (FEC)

FEC is another technique used in error-prone environments to detect and correct bit errors in bit streams. In this method, at the transmission end some additional data, known as parity bits, are enclosed in the bit stream to facilitate error recovery. The selection of the appropriate FEC codes for a particular application depends on the characteristics of the algorithm such as resource consumption, memory usage, flexibility and computational complexity.

Some of the FEC codes used extensively in literature are listed below with their key features:

1. Reed Solomon codes:
   - Simple systematic linear block error correction code.
   - With the increase in file size, channel bit-rate and error rate, the amount of overhead added increases rapidly, resulting in a massive extra bandwidth requirement.
   - Requires lot of memory for the processing. But this can be improved by means of memory management algorithms. However, that in turn results in an increase in decoding time.
   - Is capable of handling variable packet sizes with low additional overhead. Yet, this increases the computational complexity.

2. Raptor codes:
   - Systematic code.
   - With the increase of channel bit-rate, the bandwidth used for the error correcting code becomes relatively low.
   - Requires less processing time and memory.
   - Can support all the requirements of Multimedia Broadcast Multicast Service (MBMS).
3. Low density parity check (LDPC) codes:
   - Have a good block error correcting capacity.
   - Suitable for parallel implementation and use iterative coding approach.
   - Supported in Digital Video Broadcasting-Satellite-Second Generation (DVB-S2) standard.
   - Implementation is complex and for effective results the code length has to be longer.

4.3 3D Video Transmission: Example Scenarios

4.3.1 3D Video Broadcast over DVB-T

4.3.1.1 Overview of Digital Video Broadcasting

In terrestrial networks the following DVB standards will be taken into account to deliver 3DTV services:

- DVB-T: first generation terrestrial transmission standard.
- DVB-T2: second generation terrestrial transmission standard.

The DVB-T standard is the most successful digital terrestrial television standard in the world. First published in 1995, it has been adopted by more than half of all countries. One of the main advantages of DVB-T is the possibility to exploit the network in a very efficient way by using the same frequency (Single Frequency Network, SFN) in the covered area by synchronizing the transmitters, at the expense of a reduction of the available capacity with respect of the Multi Frequency Network (MFN).

Since the publication of the DVB-T standard, however, research in transmission technology has continued, and new options for modulating and error-protecting broadcast steams have been developed. Simultaneously, the demand for broadcasting frequency spectrum has increased as has the pressure to release broadcast spectrum for non-broadcast applications, making it ever more necessary to maximize spectrum efficiency. In response, the DVB Project has developed the second-generation digital terrestrial television (DVB-T2) standard. The specification, first published by the DVB Project in June 2008, was standardized by European Telecommunication Standardisations Institute (ETSI) in September 2009. Since January 2010, set-top boxes

Table 4.1 DVB-T vs DVB-T2

	DVB-T (MFN)	DVB-T2 (MFN)	DVB-T (SFN)	DVB-T2 (SFN)
Modulation	64QAM	256QAM	64QAM	256QAM
FFT size	8K	32K	8K	32K
Guard Interval	1/32	1/128	1/4	1/16
Bandwidth	Standard	Extended	Standard	Extended
Required C/N	16.5 dB	16.1 dB	16.5 dB	16.1 dB
Capacity	24.1 Mbit/s (gain = 50%)	36.1 Mbit/s	19.9 Mbit/s (gain = 70%)	34.3 Mbit/s

and TV sets have been available in the UK market immediately after the deployment of the Freeview HD DVB-T2 network. The possibility increasing the capacity in a digital terrestrial television (DTT) multiplex is one of the key benefits of the DVB-T2 standard. In comparison to the current digital terrestrial television standard, DVB-T, this second-generation standard provides a minimum increase in capacity of more than 30% in equivalent reception conditions using existing receiving antennas. Lab tests carried out at Rai CRIT indicate that the increase in capacity obtained in practice may be closer to 50% for MFN and close to 70% for SFN.

As an example of the typical configuration for an MFN network Table 4.1 confirms that 4 HD programmes can be allocated in the 8MHz bandwidth where 4–5 SD programmes are allocated with first generation DVB-T standard. In Table 4.2 the typical Modcod schemes adopted by broadcasters for fixed reception with roof-top antenna are listed, together with the bandwidth capacity and the required C/N.

DVB-T2 system encompasses a lot of modulation and coding options and many other options to choose from, for maximum flexibility. In this scenario only the modulation schemes that approach the same C/N of the DVB-T (i.e. the same coverage) are listed in Table 4.3.

Table 4.2 DVB-T capacity and required C/N

Modcod	Bit-rate [Mbit/s]	Gaussian channel C/N [dB]	Network
64QAM, rate 2/3, GI 1/32, 8k	24.1	16.5	MFN
64QAM, rate 3/4, GI 1/32, 8k	27.1	18.0	MFN
64QAM, rate 5/6, GI 1/32, 8k	30.2	19.3	MFN
64QAM, rate 2/3, GI 1/4, 8k	19.9	16.5	SFN
64QAM, rate 3/4, GI 1/4, 8k	22.4	18.0	SFN
64QAM, rate 5/6, GI 1/4, 8k	24.9	19.3	SFN

Table 4.3 DVB-T2 capacity and required C/N

Modcod	Bit-rate [Mbit/s]	Gaussian channel C/N [dB]	Network
256QAM, rate 3/5, GI 1/128, 32k	36.1	16.4	MFN
256QAM, rate 2/3, GI 1/128, 32k	40.2	18.1	MFN
256QAM, rate 3/4, GI 1/128, 32k	45.1	20.3	MFN
256QAM, rate 3/5, GI 1/16, 32k	34.3	16.4	SFN
256QAM, rate 2/3, GI 1/16, 32k	38.2	18.1	SFN
256QAM, rate 3/4, GI 1/16, 32k	42.9	20.3	SFN

DVB-T2 adds GSE (Generic Stream Encapsulation) as defined in the DVB-S2 specification, to the MPEG2-TS encapsulation. Moreover DVB-T2 supports also another important feature, the Multiple Physical Layer Pipe (MPLP): each stream (up to 255 with an arbitrary mixture of Transport Stream and Generic Streams) can have individual channel protection (code rate and constellation, Modcod). As the VCM of the DVB-S2, the Modcod cannot be driven by changes in the propagation conditions.

4.3.1.2 3DTV Transmission

The goal is to evaluate the number of 3DTV services which can be broadcast on the 8Mhz terrestrial channel. An efficient and reduced cost deployment scenario would be the transmission of 3DTV channels using Multiview Video plus Depth (MVD) and Multiview Video Coding (MVC). On the terrestrial network, although it is technically possible to broadcast Multi-layer video content adopting MPLP to have different quality layers on the same carrier, from the point of view of the service availability, the scenario is very different from the satellite, where the propagation path varies due to weather conditions (both in uplink and in downlink). Here, adopting MPLP would result in a different coverage for different streams and this condition would not be time-variant. As a conclusion, single PLP transmission will be considered in the parameters selection (i.e. modulation scheme, code rate, etc.). In Table 4.4 the number of multi-layered services as a function of the available bit-rate is summarized.

The DVB-T standard enables the transmission of one or at a maximum two basic stereoscopic layer services or just one MVD2 (two views plus relevant depth images) service. A full multi-layered 3DTV service is available only with the less protected configuration of the DVB-T2 mode taken into account.

4.3.1.3 Terrestrial System

The basic architecture of a DVB-T/T2 delivery system consists of several blocks as indicated in Figure 4.3.

Table 4.4 Multilayered 3DTV services over 8 MHz terrestrial channels

		# Layers		
		Stereoscopic	MVD2	MVD4
DVB-T	64QAM, rate 2/3, GI 1/32, 8k	1	1	–
	64QAM, rate 3/4, GI 1/32, 8k	2	1	–
	64QAM, rate 5/6, GI 1/32, 8k	2	1	–
	64QAM, rate 2/3, GI 1/4, 8k	1	1	–
	64QAM, rate 3/4, GI 1/4, 8k	1	1	–
	64QAM, rate 5/6, GI 1/4, 8k	1–2	1	–
DVB–T2	256QAM, rate 3/5, GI 1/128, 32k	2	1–2	–
	256QAM, rate 2/3, GI 1/128, 32k	3	2	1
	256QAM, rate 3/4, GI 1/128, 32k	3	2	1
	256QAM, rate 3/5, GI 1/16, 32k	2	1	–
	256QAM, rate 2/3, GI 1/16, 32k	3	2	1
	256QAM, rate 3/4, GI 1/16, 32k	3	2	1

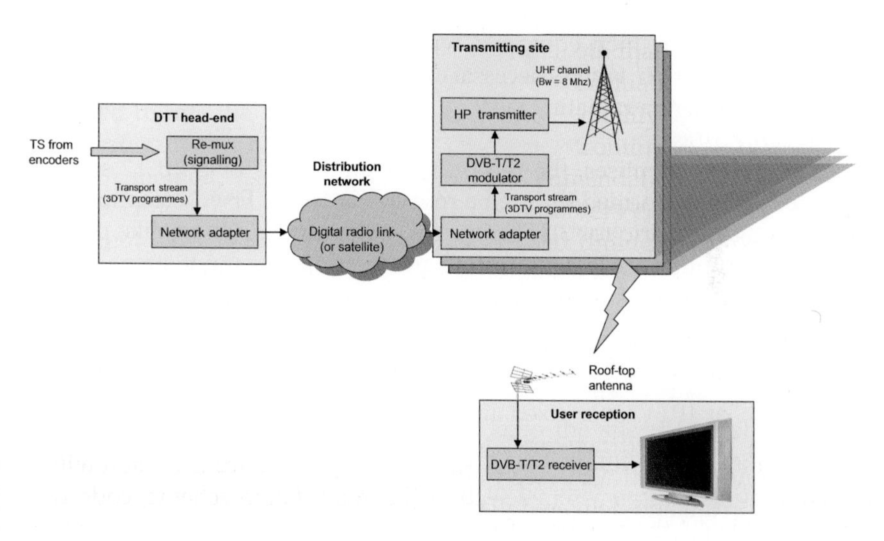

Figure 4.3 DVB-T/T2 distribution channel

DTT Head-End

In the main DTT head-end the Transport Stream from the encoding block can be processed to include additional A/V programmes, or contents such as interactivity, or simply to add additional data such as, for example, signaling or EPG information. The resulting Transport Stream is then ready for the delivery to the terrestrial transmitting sites through the distribution network.

Distribution Network

Depending on the number of transmitters the distribution network could be based on digital radio links, IP networks/optical fibres or satellite transponders. In some cases a combination of technologies is used, for example, for main transmitting sites the main feed is from a digital radio link with redundancy plus an additional back-up via satellite, while for small/minor transmitting sites only one type of network is used.

Transmitting Sites

The number of transmitting sites required may vary from a few tens up to some hundreds, depending on the area to be covered both in terms of extension and in terms of orographic condition (as an example for a country like Italy, with mountainous areas, several hundreds of transmitters are used, with some high power stations and the majority of mid-low power stations).

For a nationwide service the DTT network may use the same frequency in the whole coverage area in the so-called SFN (Single Frequency Network) mode: in this case, to avoid destructive interference in the overlap areas of two transmitters, all the transmitters must be synchronized to broadcast exactly the same bits at the same time and this functionality is achieved in the head-end by means of SFN adapters, locked to the GPS signal. Every transmitting site is then equipped with a DVB-T/T2 modulator (with an SFN adapter option where necessary) and a power amplifier feeding an appropriate antenna radiating system.

Receiving Sites

At the user's premises, the reception of the DVB-T or DVB-T2 signal is expected to be performed through a roof-top antenna. Fixed reception is the typical usage scenario for 3DTV services, other configurations like portable or mobile reception are less interesting and in addition cannot offer adequate performance (i.e. transmission capacity).

4.3.2 3D Video Streaming over IP Networks

In the case of IPTV over fixed line networks, the delivery infrastructure relies on a "managed network" to ensure bandwidth efficient delivery of vast amounts of multicast video traffic. The video data is sent as an MPEG-2 transport stream and delivered via IP Multicast in the case of Live TV or via IP Unicast in the case of Video on Demand.

4.3.2.1 Network Architecture

The IPTV broadcasting architecture relies on the following main segments:

- *Contribution/Head-End*: the part of the architecture dedicated to receiving television signals, content ingestion, processing and distribution over the IP network. It hosts Live Encoders, Centralized Video Servers, the Media Asset Management System, the Middleware Platform, Conditional Access (CAS) and DRM systems.

- *Transport/Distribution Network*: relies on IP Core Network (Optical Packet Backbone and MAN GbE). Hosts local VoD servers.
- *Access Network/Home Network*: the part of the network which connects subscribers to their immediate service provider (between DSLAM and Residential Gateway). Access Network infrastructures are based on xDSL technology.
- *Home Network*: at the customers' premises, Ethernet or Wi-Fi based, hosts the customers' CPEs (STB, TV).

The architecture is exemplified in Figure 4.4.

4.3.2.2 3DTV Transmission

We assume that it should not be required to modify the IPTV distribution network to allow transmission of 3DTV content (see D1.1.1 req. 08.01 [1]): in particular, the part of the delivery infrastructure consisting of the Distribution Network up to the DSLAM will have to be preserved. The part that is most critical in the provisioning of 3D IPTV is represented by the Access Network (from DSLAM to the customer's home), which may represent a bottleneck in terms of bandwidth.

The two 3D IPTV enabled service scenarios are "Live TV" (broadcasting events in real time to multiple users, based on Multicast mode) and "Video On Demand" (content is transmitted to a single user only when requested, based on Unicast mode).

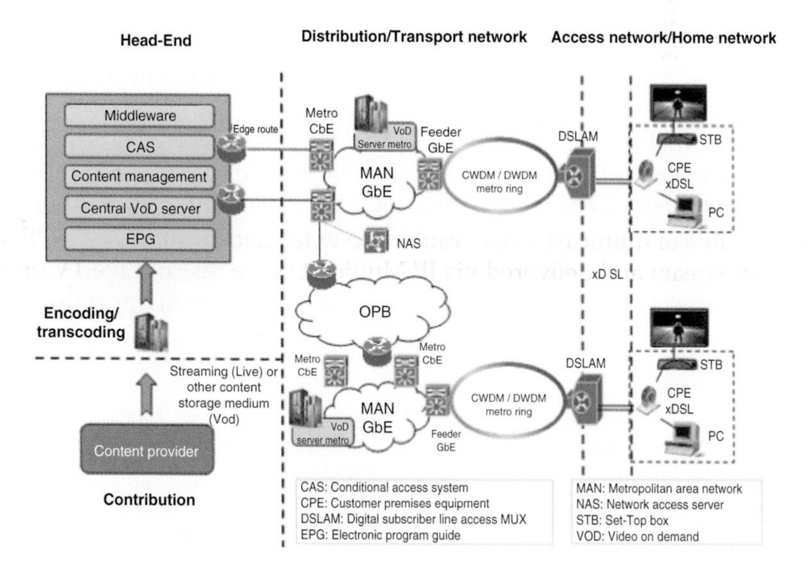

Figure 4.4 The IPTV broadcasting architecture

- ***Live TV*** – The content encoding can be performed either by the Telco (encoders located in the Telco's head-end) or directly by the Content Provider, who will also care for the correct transmission on the Telco's network. Network transmission employs a Multicast channel, which can rely either on a IP+UDP protocol stack or on IP+UDP+RTP. In the case of Telecom Italia, the IP+UDP protocol stack is employed; however, the use of RTP would not have any impact as far as the network configuration is concerned.
- ***VoD*** – Video contents are located in the central VoD server in the Telco's head-end, and replicated in the local VoD servers in the various PoPs over the metro networks. Video fruition is based on the Unicast model: the client requests the content through the RTSP protocol and the VoD server sends the content in Unicast IP+UDP.

IPTV 3D content distribution, both Live and VoD, has to take into account possible impediments related to:

- Access Network bandwidth limitations (ADSL2+, VDSL);
- CPE (STB, TV) incompatibility with the 3D service.

What is more relevant in the case of a Telco is the first point: the Access Network (the part of the network which connects subscribers to their immediate service provider) which represents the bottleneck in IPTV delivery.

- Current Access Network infrastructures are based on ADSL2+ technology (Ref. ITU G.992.5). ADSL2+ provides a maximum theoretical speed of 24 Mbit/s download and up to 3.5 Mbit/s upstream, depending on the distance from the DSLAM to the customer's home. The majority of customers' Access Networks typically rely on a bandwidth budget of a maximum 10 Mbit/s.
- In the case of IPTV, the adoption of VDSL2 technology (Ref. ITU G.993.2) for the Access Network infrastructure could increase the bandwidth capacity. VDSL2 theoretical data rates deteriorate from 250 Mbit/s at the source to 100 Mbit/s at 0.5 km and 50 Mbit/s at 1 km from the source.

With reference to data rate requirements for multi-layered transmission of 3DTV, this means that:

- ADSL2+:
  - — conventional stereo w/o depth (13 Mbit/s): can be supported in the case of optimal ADSL2+ Access Network;
  - — MVD2 (20 Mbit/s): critical support, to transmit a single 3DTV programme, it would require using the full bandwidth available in the Access Network in a theoretical optimal condition; even if technically feasible, it will be not realistic from a commercial point of view;

  - MVD4 (37 Mbit/s): not supported;
- VDSL2 (estimate based on 50 Mbit/s @ 1 km from the source):
  - conventional stereo w/o depth (13 Mbit/s): can be supported in the case of VDSL2 Access Network;
  - MVD2 (20 Mbit/s): can be supported in the case of the VDSL2 Access Network;
  - MVD4 (four views plus the relevant depth images which need an approximate bit-rate of 37 Mbit/s): critical support, to transmit a single 3DTV programme it would require using the full bandwidth available in the Access Network in a theoretically optimal condition; even if technically feasible, it will be not realistic from a commercial point of view.

In all cases the distance between the end-user device and the DSLAM plays a role, and only those in close proximity to the DSLAM will be in a better condition to receive 3DTV programmes. Therefore, for the delivery of 3DTV over IPTV, it will be important to ensure scalable transmission and bandwidth occupancy optimization by preserving, on one side, the current IPTV distribution network, and on, the other side, allowing 3DTV content fruition to users who may have different bandwidth availability.

4.3.3 3D Video Transmission over Mobile Broadband

The delivery of 3D video services, such as mobile 3DTV, 3D video streaming, and free-viewpoint video over mobile broadband networks is believed to be the next significant step in visual entertainment experience over wireless. The rapid evolution of mobile broadband technologies into an all-IP network architecture with significant improvements in data rates, coverage, and reliability would benefit from the rich content leading to new business models for the service provider and where the end-users can benefit from the improved network infrastructure to enjoy a 3D video experience on the move. 3D video services are expected to appeal to mobile subscribers by bringing 3D content to handheld devices that are equipped with glasses-free 3D display technologies, such as auto-stereoscopic displays. However, the delivery of 3D video services over wireless networks to individual users poses more challenges than conventional 2D video services. This is due to the large amount of data involved, diverse network characteristics and user terminal requirements, as well as the user's context (e.g. preferences, location). In addition, the demand for low latency high speed mobile data access is intense and the wireless data traffic is expected to continue to increase exponentially in the foreseeable future as smart phones and bandwidth-intensive wireless multimedia applications continue to proliferate in an exponential manner. This leads to straining the current cellular networks to a breaking point as the point-to-point radio link approaches its theoretical limits.

4.3.3.1 Mobile Broadband Networks

One of the most important stages in the end-to-end 3D video delivery chain over mobile broadband networks is the system used for the delivery of 3D video to the wireless subscribers. Wireless cellular technologies are continuously developing to meet the increasing demands for high data rate mobile services. The International Telecommunication Union (ITU) has been working on a new international standard called International Mobile Telecommunications-Advanced (IMT-Advanced), which is regarded as the succeeding and evolutionary version of IMT-2000, the international standard on 3G technologies and systems. Among the few technologies that are currently contending for a place in the IMT-Advanced standard include the 3rd Generation Partnership Project (3GPP) Long Term Evolution (LTE)/LTE-Advanced (LTE-A), 3GPP2 Ultra Mobile Broadband (UMB), and the Worldwide Interoperability for Microwave Access (WiMAX) based on IEEE 802.16e/m.

3GPP LTE was introduced by 3GPP in Release 8 as the next major step in the development track. The goals of LTE include higher data rates, better spectrum efficiency, reduced delays, lower cost for operators, and seamless connection to existing networks, such as GSM, CDMA, and HSPA [2]. Commercial deployments of LTE networks have already started around the end of 2009. LTE is based on an Orthogonal Frequency Division Multiple Access (OFDMA) air interface in the downlink direction and a Single Carrier-Frequency Division Multiple Access (SC-FDMA) air interface in the uplink direction. It is capable of providing considerably high peak data rates of 100 Mbps and 50 Mbps in the downlink and uplink directions, respectively in a 20 MHz channel bandwidth. LTE supports a scalable bandwidth ranging from 1.25 to 20 MHz and supports both Frequency Division Duplex (FDD) and Time Division Duplex (TDD) duplexing modes. In order to achieve high throughput and spectral efficiency, LTE uses a Multiple Input Multiple Output (MIMO) system, such as 2×2, 3×2, and 4×2 MIMO configurations in the downlink direction. The one-way latency target between the Base Station (BS) and the User Equipment (UE) terminal is set to be less than 100 msec for the control-plane and less than 5 msec for the user-plane. In addition, LTE provides IP-based traffic as well as end-to-end Quality of Service (QoS) to support multimedia services. LTE-A is being specified initially as part of Release 10 of the 3GPP specifications that will meet or exceed the requirements of the IMT-Advanced standard.

WiMAX is a broadband wireless access technology that is based on the IEEE 802.16 standard, which is also called the Wireless Metropolitan Area Network (WirelessMAN). The name "WiMAX" was created by the WiMAX Forum [3] that was formed in June 2001 to promote conformance and interoperability of the standard. It enables the delivery of last-mile wireless broadband services as an alternative to cable and Digital Subscriber Line (xDSL) wireline access technologies. The ease of deployment and the low cost of installation and maintenance are key features that make WiMAX more attractive than cable and xDSL, especially in those places where the latter two technologies cannot be used or their costs are prohibitive [4]. For example, in countries with a scarce wired infrastructure, WiMAX can become part of the broadband backbone. The WiMAX standard covers a wide range of fixed and mobile applications. The fixed version is termed IEEE 802.16d-2004, which is for fixed or slow position changing devices (e.g. personal computers, laptops). The mobile version is termed IEEE 802.16e-2005, which is an amendment to IEEE 802.16d-2004 to support mobility, targeting mobile devices travelling at speeds of up to 120 km/h (e.g. smart phones). Mobile WiMAX is based on an OFDMA air interface and supports a scalable channel bandwidth of 1.25 to 20 MHz. WiMAX is capable of offering a peak downlink data rate of up to 63 Mbps and a peak uplink data rate of up to 28 Mbps in a 10 MHz channel bandwidth with MIMO antenna techniques and flexible sub-channelization schemes. WiMAX is also competing for a place in the IMT-Advanced standard via the IEEE 802.16m, which is an amendment to IEEE 802.16-2004 and IEEE 802.16e-2005. A theoretical data rate requirement for IEEE 802.16m is a target of 1 Gbps in stationary mode and 100 Mbps in mobile mode.

4.3.3.2 3D Video Multicast/Broadcast

Unicast, multicast, and broadcast are three transmission methods that can be used for multimedia communication applications, such as 3D video communications over IP wireline and wireless networks. Unicast is a point-to-point communication between a single sender and a single receiver over a network, such as downloading, streaming media on demand, and point-to-point telephony. Multicast is the communication between a single sender and a group of receivers that are participating in a multicast session over a network, such as internet television. Multicast is more efficient than multiple unicasts in terms of network resource utilization and server complexity. Broadcast transmission connects a sender to all receivers that it can reach through the network, such as broadcasting over a wireless link or a shared Ethernet link.

The offering of 3D video services, such as mobile 3DTV to mobile subscribers using the streaming option over unicast or point-to-point connections allows an unlimited number of TV channels to be offered to a limited number of subscribers. However, it is limited both from a cost and technical viewpoint when delivering such services to many subscribers at the same time. This is due to the fact that if numerous recipients try to access the service at the same time (e.g. during a live sports event), the network is going to saturate and the operator will not be able to offer the service adequately with the inevitable decline in QoS. Hence, better to limit the audience and prevent a mass-market deployment of 3D video services.

Using an overlay broadcast access network technology is an alternative approach to offer mobile TV and mobile 3DTV services for the mass market. Globally, the research and development of 3D content delivery in the form of mobile 3DTV over wireless networks have also been actively conducted based on mobile broadcasting technologies, such as Terrestrial-Digital Multimedia Broadcasting (T-DMB), based on the Eureka-147 Digital Audio Broadcasting (DAB) standard and Digital Video Broadcasting-Handheld (DVB-H) systems [4]. However, each of the aforementioned broadcast access network technologies would require an independent network infrastructure to be built, implying additional deployment and operational costs for the service provider.

The other alternative approach is to take advantage of cellular broadcast techniques, whereby utilizing such techniques would enable cellular network operators to take advantage of their already-deployed network infrastructure to alleviate the bandwidth impacts of unicast for at least some types of content, such as mobile 3DTV. Such techniques are supported by the emerging mobile broadband wireless access network technologies, such as WiMAX Multicast and Broadcast Service (MBS) and LTE Multimedia Broadcast Multicast Service (MBMS), which are considered one of the viable wireless networking technologies to meet the uptake of mobile TV and mobile 3DTV services. However, the solutions that are based on utilizing cellular broadcast techniques allow a limited number of TV channels to be offered to an unlimited number of subscribers.

In considering the aforementioned unicast, overlay broadcast, and cellular broadcast transmission approaches, it becomes evident that it is essential to ensure that the radio resources are not wasted by multicasting/broadcasting channels that only a few people watch, and that all users can receive the channels they want to watch. One possible solution would be the combination of the different transmission approaches into a hybrid approach to meet the consumers' and the operators' interests [5]. For example, the most popular channels can be broadcast using the cellular broadcast techniques (e.g. WiMAX MBS, LTE MBMS) or an overlay broadcast network if available. The secondary channels can be offered using multicast or unicast depending on the demand in each particular cell or area. Interactive services and niche channels can be selectively offered to the subscribers over unicast links.

4.3.3.3 Cellular Multicast/Broadcast Services

MBMS enables an efficient transmission mechanism in the downlink direction for multimedia content, such as multicasting/broadcasting 3DTV content to large-scale user communities. MBMS was initially introduced in 3GPP Release 6 for UMTS; however, only a small number of operators have conducted MBMS service trials so far [6]. For 3GPP LTE, in order to reduce the bandwidth impacts of unicast video traffic for at least some types of the content, one of the key features in LTE Release 9 is the Evolved-MBMS (E-MBMS). In the downlink direction, the OFDMA-based LTE air interface now offers a characteristic that is better suited to such services, such as cell-edge spectrum efficiency in urban or suburban environments, exploiting the special features of E-MBMS. In E-MBMS operation, there are two scenarios of the single-cell broadcast and the Multimedia Broadcast Single Frequency Network (MBSFN), where both typically use the PMP mode of transmission. Therefore, UE feedback, such as Acknowledgement (ACK)/Negative ACK (NACK) and Channel Quality Information (CQI) are not utilized as is the case in point-to-point mode of transmission. However, one solution to this would be the utilization of aggregate statistical ACK/NACK and CQI information for link adaptation and retransmissions.

For MBSFN operation, the MBMS data is transmitted simultaneously from a number of time-synchronized evolved NodeBs (eNBs) using the same resource block/s. In this case, the UE receiver can observe multiple versions of the same transmitted signal with different propagation delays. As a result, the UE can combine the transmissions from the different eNBs, which can greatly enhance the Signal-to-Interference-and-Noise Ratio (SINR) in comparison to the non-SFN operation. The advantage of the MBSFN operation is especially evident at the cell edge, where the MBMS transmissions that would otherwise have constituted inter-cell interference are now translated into useful signal energy; hence, the received signal power is increased at the same time as the interference power is largely removed. In MBSFN, the used Cyclic Prefix (CP) is slightly longer. Inter Symbol Interference (ISI) should not be an issue if the time synchronization of the different eNB transmissions is sufficiently tight for each transmission to arrive at the UE within the CP at the start of the symbol [7]. Figure 4.5 illustrates the overall user-plane architecture for the MBSFN operation and MBMS content synchronization.

The eBM-SC is the MBMS traffic source and the E-MBMS gateway is responsible for distributing the MBMS traffic to the different eNBs of the MBSFN area, where the IP multicast may be used. Header compression for MBMS services can be performed by the E-MBMS gateway, as 3GPP has currently assumed.

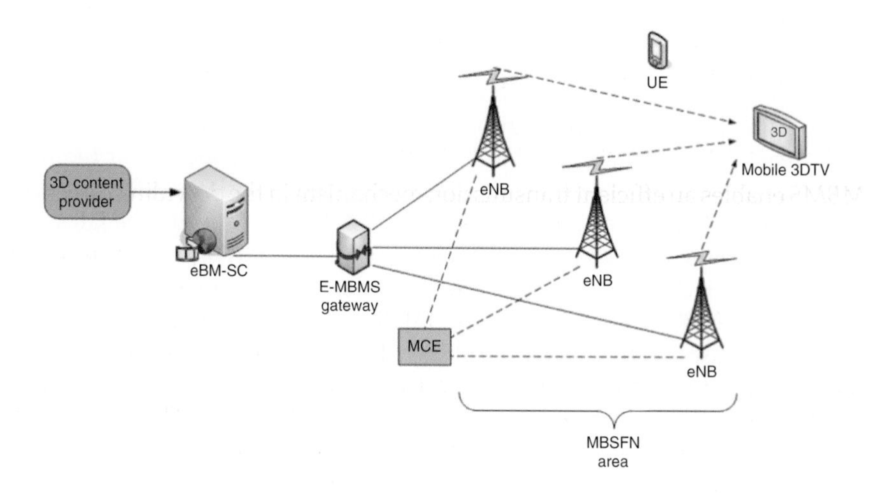

Figure 4.5 Overall user-plane architecture for MBSFN operation

In order to ensure that the same MBMS content is transmitted from all the eNBs of the MBSFN area, a SYNC protocol has been defined by 3GPP between the E-MBMS gateway and the eNBs. 3GPP has also defined the MBMS Coordination Entity (MCE), which is a control-plane entity ensuring that the same resource block is allocated for a given service across all the eNBs of a given MBSFN area. In addition, it ensures that the Radio Link Control (RLC) and Medium Access Control (MAC) sublayers at the eNBs are configured appropriately for the MBSFN operation.

WiMAX MBS provides an efficient transmission method in the downlink (DL) direction for the concurrent transport of data common to a group of Mobile Stations (MSs) through a shared radio resource and using a common Connection IDentifier (CID) [8]. The area where the MBS service is offered is called an MBS Zone, which may consist of one or more than one BS. Each MBS Zone is identified by a unique MBS Zone ID that is not reused across any two adjacent MBS Zones. In the single-BS provisioning of MBS, the BS provides the MSs with MBS using any multicast CID value locally within its coverage area and independently of other BSs. In the multi-BS provisioning of MBS, all BSs within the same MBS Zone use the same CIDs and Security Associations (SAs) to transmit the content of a certain common MBS.

Coordination between these BSs allows an MS to continue to receive MBS transmissions from any BS within the MBS Zone, regardless of the MSs operating mode (e.g. Normal, Idle) and without the need to re-register for the BS. Optionally, MBS transmissions may be synchronized across all BSs within an MBS Zone, enabling an MS to receive the MBS transmissions from multiple BSs using macro-diversity, hence, improving the reception reliability. The MBS service can be constructed by either constructing a separate MBS Region in the downlink subframe along with other unicast services, i.e. integrated or embedded MBS, or the whole frame can be dedicated to MBS, i.e. downlink only for a standalone broadcast service. In addition, it is also possible to construct multiple MBS Regions. This is illustrated in Figure 4.6 that shows the downlink subframe construction when a mix of unicast and multicast/broadcast services are supported.

For each MBS Region, there is one MBS MAP that is located at the first subchannel and the first OFDMA symbol of the associated MBS Region. The MBS MAP contains multiple MAP DATA IEs that specify the CID, the location, and the PHY configuration of one MBS burst. In the DL-MAP, there is one MBS MAP Information Element (IE) descriptor per MBS Region that specifies the MBS Region PHY configuration and defines the location of each MBS Region via the OFDMA Symbol Offset parameter. The MS accesses the DL-MAP to initially identify the MBS Regions and the locations of the associated MBS MAPs in each region. Then the MS can subsequently read the MBS MAPs without reference to the DL-MAP unless synchronization to the MBS MAP is lost.

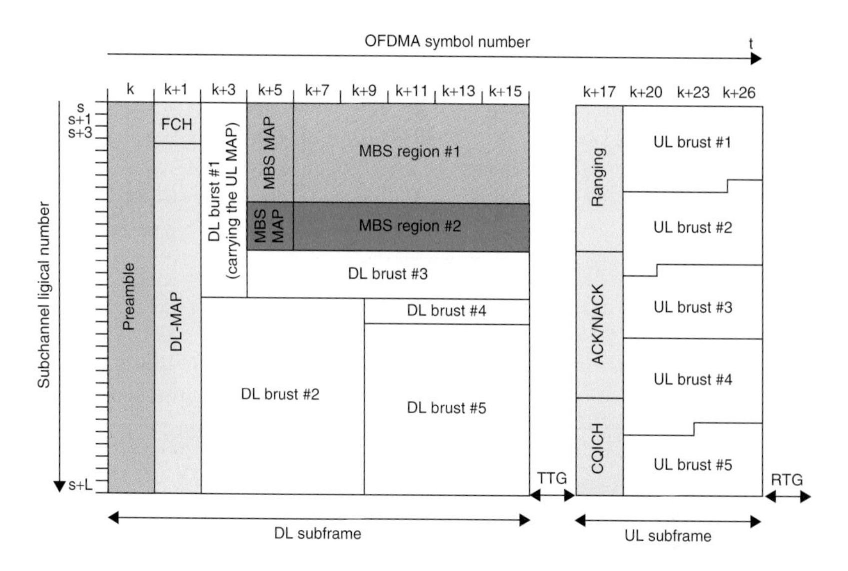

Figure 4.6 Construction of MBS Regions in an OFDMA frame in TDD mode

4.3.3.4 Transmission Protocols for 3D Video Services

The transmission of video bit streams that are encoded using the H.264/MPEG-4 AVC video coding standard [9] over the Real-time Transport Protocol (RTP) [10, 11] is standardized by the IETF in RFC 3984 [12]. It describes an RTP Payload format for the ITU-T Recommendation H.264 video codec. In this format, one or more Network Abstraction Layer (NAL) units that are produced by an H.264 video encoder can be packetized in each RTP payload. For 3D video transmission, this RTP payload format is suitable for simulcast coded 3D video representation formats, such as two-channel stereo video encoded independently using H.264/AVC and colour-plus-depth. In [13], an Internet Draft that describes an RTP Payload format for the transmission of 3D video bit streams encoded using the Multi-view Video Coding (MVC) [9] mode of the ITU-T Recommendation H.264 video codec over RTP is described. This RTP Payload format allows the packetization of one or more NAL units that are produced by the video encoder in each RTP payload. For example, instead of sending the MVC stream as a single H.264/AVC stream over RTP, the stream is divided into two parts where NAL units are transmitted over two different RTP port pairs as if they are separate H.264/AVC streams. The RTP packets are then transmitted over UDP [14] and IP [15]. Figure 4.7 and Figure 4.8 illustrate the user plane protocol stack for 3D video delivery over LTE and mobile WiMAX networks, respectively, starting at the 3D content server and ending at a UE or an MS in a cell.

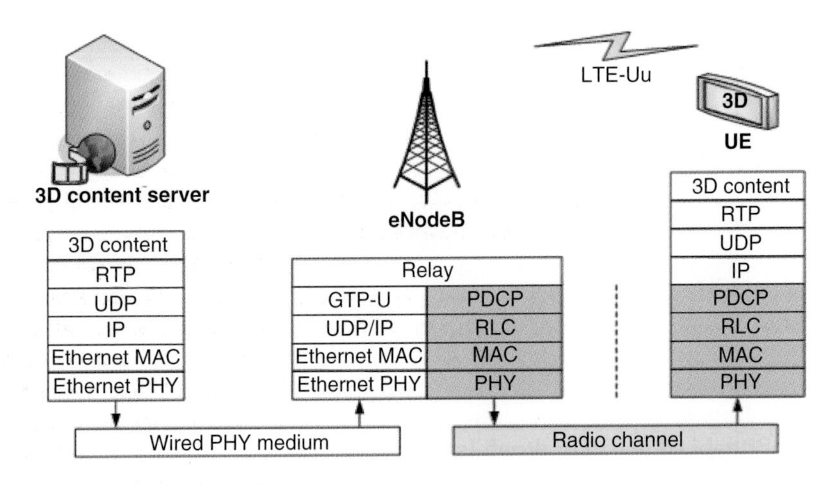

Figure 4.7 User plane protocol stack for 3D video delivery over LTE

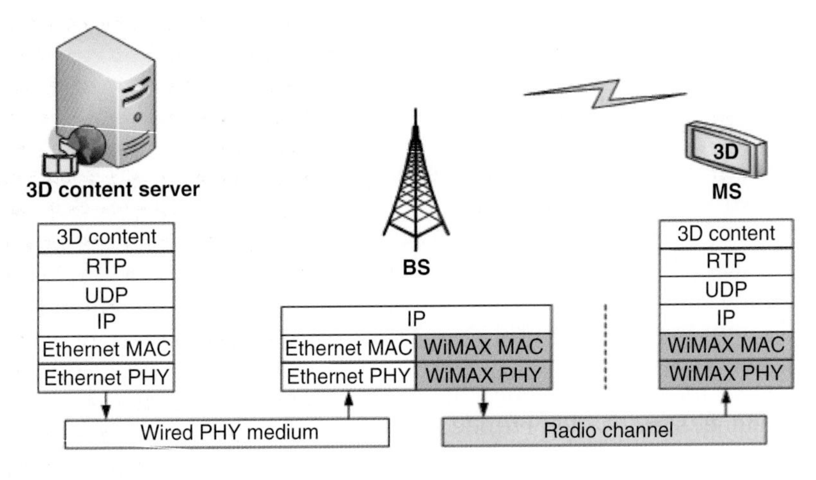

Figure 4.8 User plane protocol stack for 3D video delivery over mobile WiMAX

4.3.3.5 Support of Heterogeneous Devices

An important requirement that should be taken into consideration in the development and deployment of 3D video services, such as mobile 3DTV over mobile broadband networks is the support of a variety of mobile device capabilities. The aim is to ensure that legacy mobile devices without 3D capabilities can consume the same content in conventional 2D. This can be achieved either in the application layer or in the lower layers, such as the data link layer and/or the physical layer of the mobile broadband technologys radio interface protocol stack.

At the application layer, several ways can be used depending on the utilized 3D video representation format and the utilized coding scheme. For example, in the case of two-channel format, this can be achieved by independently encoding the left and right views using existing 2D video encoders, such as H.264/MPEG-4 AVC. Similarly for the colour-plus-depth format, the colour and the depth map components are encoded independently using conventional 2D video encoders. If the MVC mode of the H.264/MPEG-4 AVC video coding standard is used to encode the left and right views of the two-channel format, the base view (e.g. the left view) is encoded conforming to at least one of the profiles of H.264/MPEG-4 AVC and the non-base view (e.g. the right view) is encoded with respect to the left view. In order to differentiate between the left view and the dependent right view, MVC Network Abstraction Layer (NAL) units of type (Type 20) are used, which are coded slices of non-base views, allowing the legacy H.264/MPEG-4 AVC decoders to ignore this data and extract and decode a 2D version of the content easily. H.264/MVC-compliant decoders can gain the full benefits of the additionally coded view and can decode the complete 3D video bit stream including the dependent view. The support of heterogeneous devices at the application layer has the advantage of being a straightforward approach that is easy to implement. However, the disadvantage is that it will be required to receive all the data up to the application layer in the mobile device, although the data corresponding to one of the views is not necessary or required. Media Aware Network Elements (MANEs) can be employed at base stations to truncate and/or manipulate the RTP streams.

At the data link layer, for example, at the BSs MAC layer in a WiMAX network, the incoming IP packets carrying the left-view and the right-view data in the case of two-channel stereo format or carrying the colour and depth map data in the case of colour-plus-depth format can be classified into different MAC transport connections. The client software at the mobile device can include a video format selector that determines how many transport connections can be utilized simultaneously. Therefore, according to the 3D capabilities of the device, it is possible to determine in which video format (e.g. 2D, 3D) the content can be viewed. The determination of the transport connections indicates the WiMAX mobile device MAC/PHY layers to process only those MBS Regions' data that are associated with the video format from the corresponding MBS Regions at the PHY layer.

4.3.3.6 Radio Resource Efficient Transmission

Considering the fact that 3D video services, such as mobile 3DTV, demand higher radio system resources than conventional 2D video services, the efficient utilization of the available radio resources is one of the key issues to be addressed when delivering such services over mobile broadband networks. We will now discuss mobile broadband network capacity calculations.

Table 4.5 Reference 3GPP LTE downlink system parameters

Parameter	Value
Carrier frequency [GHz]	2
Channel bandwidth [MHz]	10
Duplexing mode	FDD
Subcarrier spacing [kHz]	15
Number of subcarriers	600
Number of subcarriers per PRB	12
Number of PRBs	50
PRB bandwidth [kHz]	180
TTI duration [ms]	1
Slot duration [ms]	0.5
Number of OFDM symbols per slot	7

3GPP LTE Capacity. In an OFDMA-based 3GPP LTE downlink, a Physical Resource Block (PRB) is the smallest element of radio resource allocation that has both a time and a frequency dimension. In the frequency domain, a PRB consists of 12 subcarriers with 15 kHz subcarrier spacing and an 180 kHz bandwidth. In the time domain, a PRB consists of one time slot of 0.5 ms duration, during which 7 or 6 OFDM symbols are transmitted, depending on whether short or long CP has been utilized, respectively. During each scheduling interval, the smallest radio resource unit that a scheduler can allocate to a user is a scheduling block (SB), which consists of two consecutive PRBs that span a subframe or a Transmission Time Interval (TTI) duration of 1 ms and a bandwidth of 180 kHz. In the following, capacity calculations are provided for a reference 3GPP LTE system. The reference systems parameters are listed in Table 4.5.

For a user i at time t, the data rate achieved ($DataRate_i(t)$) at time t on two consecutive PRBs or a scheduling block, which spans a subframe or a TTI duration of 1 ms and a bandwidth of 180 kHz, is calculated according to the following equation:

$$DataRate_i(t) = N_{bits/symbol} \times N_{symbols/slot} \times N_{subcarriers/PRB} \times N_{PRBs/TTI} \times CC_{rate} \times \frac{1}{TTI_{duration}} \tag{4.1}$$

Where ($N_{bits/symbol}$) is the number of bits per OFDM symbol on a subcarrier within a PRB, which can be determined using the approach proposed in [16]. ($N_{symbols/slot}$) is the number of OFDM symbols per 0.5 ms slot, ($N_{subcarriers/PRB}$) is the number of subcarriers per PRB, ($N_{PRBs/TTI}$) is the number of PRBs per subframe or a TTI duration of 1 ms, (CC_{rate}) is the channel coding rate, and ($TTI_{duration}$) is the subframe duration of 1 ms.

Table 4.6 Peak downlink data rates per two PRBs for various MCSs

Modulation type	Number of bits per symbol	Channel-coding rate	Peak data rate per two PRBs [Mbps]
QPSK	2	1/2	0.168
		3/4	0.252
16-QAM	4	1/2	0.336
		2/3	0.448
		3/4	0.504
64-QAM	6	1/2	0.504
		2/3	0.672
		3/4	0.756
		5/6	0.84

Table 4.6 lists the theoretical peak data rates for a user i at time t per scheduling block in the downlink direction for various Modulation and Coding Schemes (MCSs).

In WiMAX OFDMA PHY, a slot is the minimum possible data allocation unit that requires both a time and a subchannel dimension for completeness. An OFDMA slot definition depends on the OFDMA symbol structure, which varies for the uplink and downlink directions, for full usage of subchannels (FUSC) and partial usage of subchannels (PUSC), and for the distributed subcarrier permutations and the adjacent subcarrier permutation. For downlink FUSC and downlink optional FUSC using the distributed subcarrier permutation, one slot is one subchannel by one OFDMA symbol. For downlink PUSC using the distributed subcarrier permutation, one slot is one subchannel by two OFDMA symbols. In the following, capacity calculations are provided for a reference mobile WiMAX system, whose parameters are listed in Table 4.7.

The OFDMA parameters for the reference system are listed in Table 4.8.

Table 4.7 Reference WiMAX system parameters

Parameter	Value
PHY layer interface	WirelessMAN-OFDMA
Carrier frequency [GHz]	2.3
Channel bandwidth [MHz]	8.75
Duplexing mode	TDD
DL:UL ratio	2:1
TDD frame length [ms]	5
TTG / RTG [μs]	121.2 / 40.4
Subcarrier permutation	PUSC

Table 4.8 OFDMA parameters

Parameter	Value
Channel bandwidth [MHz]	8.75
Sampling factor	8/7
Sampling frequency (F_s) [MHz]	10
Sample time ($1/F_s$) [ns]	100
FFT size (N_{FFT})	1024
Subcarrier spacing (Δf) [kHz]	9.76
Useful symbol time ($T_B = 1/\Delta f$) [μs]	102.4
Guard time or cyclic prefix ($T_G = T_B/8$) [μs]	12.8
OFDMA symbol time ($T_S = T_B + T_G$) [μs]	115.2

At 8.75 MHz channel bandwidth, the OFDMA symbol time is equal to 115.2 s; therefore, there are 43.4 symbols in the 5 ms TDD frame. Of these 43.4 symbols, 1.4 symbols are used for the TTG and RTG, leaving 42 symbols. If n of these 42 symbols are used for the DL subframe, then $(42 - n)$ symbols are available for the UL subframe. Since PUSC is used, one DL slot is one subchannel by two OFDMA symbols and one UL slot is one subchannel by three OFDMA symbols.

Therefore, the available 42 symbols can be divided such that $(42 - n)$ is a multiple of 3 and n is of the form $(2k + 1)$. For a DL:UL ratio of 2:1, these considerations would result in a DL subframe of 27 symbols and an UL subframe of 15 symbols. In this case, the DL subframe will consist of a total of 13×30 or 390 slots without preamble (one symbol column for preamble). The UL subframe will consist of 4×35 or 140 slots excluding one slot column for ranging, ACK/NACK, and CQICH. Table 4.9 lists the OFDMA slot capacity (spectral efficiency) and the theoretical peak data rates in the downlink direction for various MCSs, or in other words, for different numbers of data bits per slot.

Dynamic radio resource allocation or packet scheduling is an algorithm or a set of algorithms that are implemented and executed at the MAC layer of BSs within a mobile broadband network. Such algorithms have the main task of allocating and de-allocating shared radio resources among users to data- and control-plane packets at each time instant in the uplink and downlink directions. The allocated and de-allocated resources can also include buffer and processing resources [7]. The aim of packet scheduling algorithms is to achieve as efficient a resource utilization as possible, while taking into account the QoS requirements of the scheduled flows. Packet scheduling algorithms perform user selection, radio resource allocation, and selection of radio bearers whose packets are to be scheduled. A packet scheduling decision typically takes into account the QoS requirements that are associated with the radio bearers, the channel quality information fed back by the users, the queuing delay of the packets, fairness indicators, buffer status, and the interference situation, etc. In addition, it may also take into account

Table 4.9 Spectral efficiency and peak downlink data rates for various MCSs

Modulation type	Number of bits per symbol	Channel-coding rate	Number of data bits per slot	Peak data rate [Mbps]
QPSK	2	1/8	12	0.936
		1/4	24	1.872
		1/2	48	3.744
		3/4	72	5.616
16-QAM	4	1/2	96	7.488
		2/3	128	9.984
		3/4	144	11.232
64-QAM	6	1/2	144	11.232
		2/3	192	14.976
		3/4	216	16.848
		5/6	240	18.72

restrictions or preferences on some of the available radio resources due to inter-cell interference coordination considerations. The scheduling decision covers not only the radio resource assignment but also which modulation and coding scheme to use and whether or not to apply MIMO or beamforming. Packet scheduling is performed once per certain time interval depending on the mobile broadband technology. For example, it is performed once per subframe or a TTI duration of 1 ms in a 3GPP LTE system and once every 5 ms in a WiMAX OFDMA TDD radio frame with a 5 ms duration.

When a wireless subscriber has been scheduled for transmission by the packet scheduler, a closely related aspect to packet scheduling is link adaptation, which deals with adjusting the transmission parameters in order to match the variations of the radio link quality over time. The modulation type (e.g. QPSK, 16-QAM, 64-QAM) and the channel coding rate (e.g. 1/2, 2/3, 3/4), i.e. the MCS used for a scheduled radio link, can be adjusted depending on the instantaneous radio channel condition such that an MCS is chosen that is most suitable for the radio channel condition at each instant, which leads to a highest average data rate while fulfilling the quality requirements. This allows high throughput, which implies low latencies and high SNR that implies low bit error rates. For example, using a robust modulation type, such as QPSK at low SNR (e.g. if the mobile device is far from the BS) and by using a spectrally-efficient modulation type, such as 64-QAM where the SNR is highest (e.g. a mobile device is close to the BS).

4.3.3.7 Transmission Techniques

Colour-plus-depth, Layered Depth Video (LDV), Depth Enhanced Stereo (DES), and Multi-View plus Depth map (MVD) 3D video representation formats all have a multi-layered bit stream structure. This layered structure may comprise parts that are perceptually more important than others or the decoding of some parts of the multi-layered 3D video bit stream is only possible if the corresponding more important parts are received correctly (e.g. base view or base quality layer). The effective exploitation of such importance and redundancy characteristics in 3D video representation formats and their associated coding approaches can be used to improve the system's performance in terms of spectral efficiency and capacity. At the physical layer of OFDMA-based mobile broadband systems, such as WiMAX and LTE, the radio transmission properties are described using a set of parameters, such as the modulation type (e.g. QPSK, 16-QAM, 64-QAM), the FEC code type (e.g. CTC, BTC), and the channel-coding rate (e.g. 1/2, 2/3, 3/4, 5/6), i.e. the MCS. The base station can utilize different MCSs to achieve the best trade-off between the spectrum efficiency and the resulting application level throughput. As a result, different transmission techniques can be adopted for the delivery of a multi-layered 3D video bit stream over a mobile broadband network.

In the first technique, referred to here as the robust transmission technique, the different components of the multi-layered 3D video bit stream are transmitted using a robust modulation type and a low channel-coding rate, such as QPSK 1/2. In this case, the service quality is ensured in an average RF environment, such as the case in mobile broadcast systems, where the transmission is typically designed to serve the worst-case user. However, the utilization of this technique would exhaust the available radio resources and hence would affect the number of offered mobile 3DTV channels and/or other services (e.g. VoIP, data, etc.). The second technique, referred to here as the radio resource efficient transmission technique, aims to offer a 3D video service as efficiently as possible. The different components of the multi-layered 3D video bit stream are transmitted using an efficient modulation type and a high channel-coding rate, such as 64-QAM 3/4, if allowed by the RF conditions of the target coverage area of the BS. However, in this case, the area over which the 3D video service can be offered will be limited and when entering difficult or bad reception conditions, the subscribers may experience sudden service interruptions instead of soft degradation in 3D video quality perception.

In the third technique, referred to here as the optimized transmission technique, the first and the second transmission techniques are combined into a unified technique so that the limitations of each are overcome by the primary advantage of the other. In this case, the coverage area of the BS can be divided into several overlapping transmission regions, where the BS multicasts/broadcasts the different components of the multi-layered 3D video bit stream in parallel using different spectrally-efficient MCSs. For example, the baseline component, which can be a two-channel stereo video encoded jointly using H.264/MVC is transmitted using a robust MCS (e.g. QPSK 1/2). This allows the subscribers with 3D-capable devices to receive

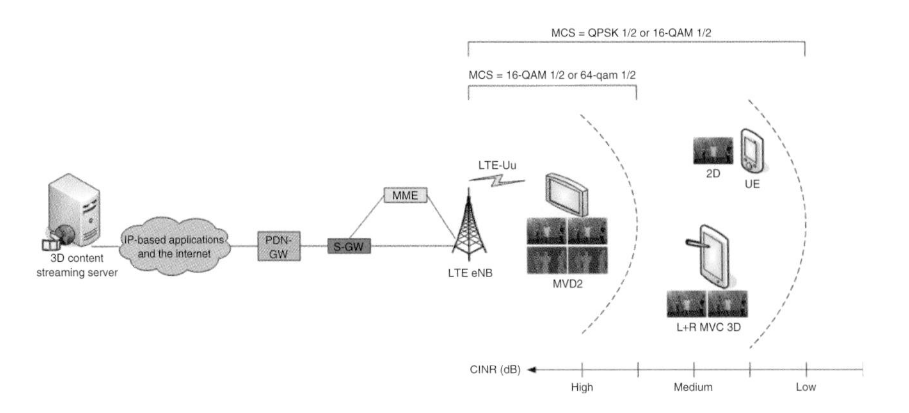

Figure 4.9 Example MVD2 transmission over an LTE network

two-channel stereoscopic 3D video format over the whole coverage area of the BS. In addition, the subscribers with legacy devices can receive the conventional 2D video version of the same content. Additional components (e.g. disparity maps) are transmitted using an efficient MCS (e.g. 64-QAM 1/2 or 3/4), allowing the subscribers that are receiving good or high signal qualities to receive the additional components adequately in addition to the baseline component. In this case, those subscribers can display an enhanced 3D video format, such as MVD2, in which there are two views and two associated depth maps; hence, enjoying better perceived video quality.

An example of the optimized transmission technique is illustrated in Figure 4.9, which shows the parallel transmission of MVD2 multi-layered 3D video bit stream over an LTE network using different spectrally-efficient MCSs.

In order to improve the reliability of the reception of the transmitted multi-layered 3D video bit streams, especially at the cell edge, while also considering the mobility aspect of the UEs, the MBSFN concept can be utilized. This will also enhance the performance in terms of avoiding the situation of unstable video decoding result between 2D and 3D due to channel quality fluctuations. This is illustrated in Figure 4.10, which shows the multi-eNB provisioning of 3D service multicast/broadcast.

In this case, the regions of the multicast/broadcast transmissions are synchronized across all the eNBs within an MBSFN area, i.e. the same information is transmitted using the same time-frequency resources from multiple time-synchronized eNBs. This enables a UE to receive the MBMS transmissions from multiple eNBs using macro-diversity, where the resultant signal level at the UE is obtained from the sum of individual signals from all the eNBs in the multi-eNB MBSFN area.

Depending on the operator's 3D video service business model, a choice can be made between the afore-mentioned transmission techniques, where two cases would be mainly considered. In the first case, the business model's

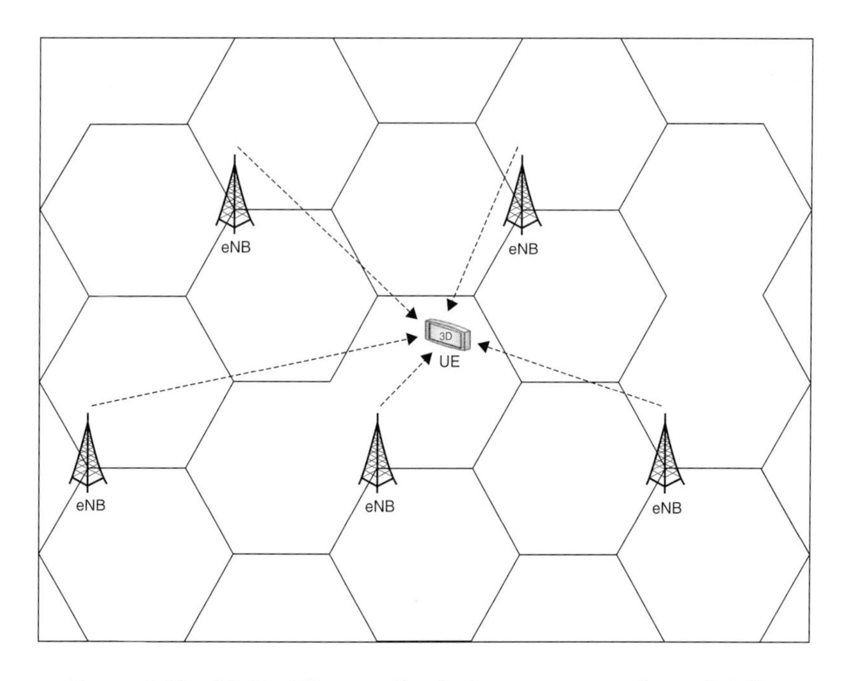

Figure 4.10 Multi-eNB reception to improve reception reliability

aim is to maximize the service quality of the offered 3D video service, so that subscribers at the cell edge would be able to receive the different components of the multi-layered 3D video bit stream adequately. This can be realized by utilizing the robust transmission technique.

In the second case, the business model's aim is to offer the 3D video service as efficiently as possible, thereby consuming fewer resources per TV channel, and hence the operator would be able to offer more TV channels and/or other services; hence, maximizing the number of subscribers per BS and generating more revenue. This can be realized by utilizing the radio resource efficient transmission technique. Considering the two business model cases and the fact that it is essential to ensure that the network is well designed in order to achieve a good compromise between service quality and the radio resource efficiency in a system that has limited resources in comparison to wired networks, the use of the optimized transmission technique would be able to provide such a compromise. In addition, this transmission technique allows for the following features: (1) video format scalability, ranging from, for example, conventional 2D video, to two-channel stereoscopic 3D video, and to MVD2 3D video. This enables the content to be displayed with the best possible quality based on the subscriber's device characteristics (e.g. display type, size), the experienced propagation conditions, and the offered package (e.g. free-to-air, pay-per-view, subscription); and (2) radio resource efficiency: reducing the amount of required radio resources to transmit one

3DTV channel, thereby allowing more TV channels to be offered or more radio resources to be available for other services.

4.4 Conclusion

Regardless of the network bandwidth limitations and the varying network conditions, delivering 3D video content to end users, while maintaining the Quality of Experience at an acceptable level, is a paramount task. This induces a plethora of research opportunities to explore efficient mechanisms for 3D video transmission across networks. One of the major challenges in real-time video delivery is to combat packet losses under noisy channel conditions. Retransmission using Automatic Repeat reQuest (ARQ) is not a feasible solution due to strict delay constraints. One of the conventional methods for reliable video multicast is to increase redundancies in the physical layer and support channel coding to recover bit errors. However, as the bandwidth requirements to transmit 3D video content increase, it is essential to investigate the most efficient means of adding redundancies. With these redundancies, the effective bit-rate increases rapidly and it may not be possible for some of the mediums that are considered here (e.g. WiMAX) to support this. This opens new research opportunities to investigate a cross-layer approach to use in 3D video transmission to efficiently distribute redundancies to maximize the end user's QoE with a limited bit budget.

References

[1] http://www.muscade.eu, 'ICT MUSCADE Project (Multimedia Scalable 3D for Europe)'.

[2] Holma, H. and Toskala, A. (2007) *WCDMA for UMTS-HSPA Evolution and LTE*, John Wiley & Sons, Ltd, Chichester.

[3] WiMAX Forum, http://www.wimaxforum.org/home.

[4] Forsman, J., Keene, I., Tratz-Ryan, B. and Simpson, R. (2004) 'Market opportunities for WiMAX take shape', Gartner Inc., December, http://www.gartner.com.

[5] Hur, N., Lee, H., Lee, G.W., Lee, S.J., Gotchev, A. and Park, S-I. (2011) '3DTV broadcasting and distribution systems', *IEEE Transactions on Broadcasting*, **57** (2), 395–407.

[6] Teleca (2011) 'Increasing broadcast and multicast service capacity and quality using LTE and MBMS. Solution area: e-MBMS in LTE', Tech. Rep., http://www.teleca.com, February.

[7] Third Generation Partnership Project (3GPP) (2011) 'Evolved universal terrestrial radio access (E-UTRA) and evolved universal terrestrial radio access network (E-UTRAN); overall description; stage 2', Tech. Rep. TS 36.300, V10.4.0 (Release 10), June.

[8] IEEE Std 802.16–2009 (2009) 'IEEE standard for local and metropolitan area networks part 16: Air interface for broadband wireless access systems', Tech. Rep., Institute of Electrical and Electronics Engineers.

[9] TELECOMMUNICATION STANDARDIZATION SECTOR OF ITU (2012) 'H.264: Advanced video coding for generic audiovisual services', Tech. Rep., International Telecommunications Union, January.

[10] Schulzrinne, H., Casner, S., Frederick, R. and Jacobson, V. (1996) 'RTP: A transport protocol for real-time applications', Tech. Rep. RFC 1889, Internet Engineering Task Force (IETF), January.

[11] Schulzrinne, H., Casner, S., Frederick, R. and Jacobson, V. (2003) ''RTP: A transport protocol for real-time applications'', Tech. Rep. RFC 3550, Internet Engineering Task Force (IETF), July.

[12] Wenger, S., Hannuksela, M.M., Stockhammer, T., Westerlund, M. and Singer, D. (2005) 'RTP payload format for H.264 video', Tech. Rep. RFC 3984, Internet Engineering Task Force (IETF), February.

[13] Wang, Y.-K., Schierl, T., Skupin, R. and Yue, P. (2012) 'RTP payload format for MVC video', Tech. Rep., Internet Engineering Task Force (IETF), June.

[14] Postel, J. (1980) 'User datagram protocol', Tech. Rep. RFC 768, Internet Engineering Task Force (IETF), August.

[15] Postel, J. (1981) 'Internet protocol', Tech. Rep. RFC 791, Internet Engineering Task Force (IETF), September.

[16] Xiaoxin Qiu and Chawla, K. (1999) 'On the performance of adaptive modulation in cellular systems', *IEEE Transactions on Communications*, **6**, 884–895.

5

Rendering, Adaptation and 3D Displays

Visualization of 3D media can be realized in a variety of ways, in contrast to the traditional 2D media. A number of 3D display technologies are available, all of which differ in terms of applied image processing and rendering techniques and the requirements for input media size and format. Thus, the mode of displaying the reconstructed 3D scene also differs. A standard 3D display on the market can display two fixed views at a time, regardless of the viewing angle from the screen, whereas multi-view displays usually have the capability of displaying more than two views simultaneously, so that different viewers looking from different viewing angles would see different pictures. This necessitates a series of specific processing tasks within the display or the display driver dedicated to the 3D mode of viewing.

This chapter explains the stages of 3D video rendering and outlines the details of various 3D display technologies. Adaptation, on the other hand, is another subject that needs special attention in the context of 3D and multi-view applications. This chapter also emphasizes several elements of various adaptation schemes devised for 3D media systems, while outlining the inherent differences from the 2D media adaptation.

5.1 Why Rendering?

Rendering refers to the process of generating a display image from a binary model. A binary scene model contains objects with a particular data structure, such as the geometry, texture and lighting comprising the description of the scene. This data is processed within the rendering engine to output a digital image, or graphics. Rendering is an inherent part of the majority

3DTV: Processing and Transmission of 3D Video Signals, First Edition.
Anil Fernando, Stewart T. Worrall and Erhan Ekmekcioğlu.

of multimedia applications, such as movies, television, streaming media, video games, etc. In its simplest form, a video renderer engine integrated to a media player carries out the task of displaying a reconstructed two-dimensional image sequence in real time. Nevertheless, in the wider sense, video rendering is usually composed of a group of computation-intensive and memory-demanding tasks, where purpose-built hardware structures called Graphical Processing Units (GPU) have been designed and integrated within modern computing architectures. Rendering algorithms can be grouped under real-time or non-real-time processes, depending upon the type of application. Especially in 3D graphics case, processing can be done offline, such as in movie post-production. On the other hand, as in video games, the rendering engine must operate in real time.

From the point of view of modern 3D video systems driven by multi-view videos, virtual view synthesis is a crucial part of the video rendering engine, affecting the perceptual quality of the 3D video significantly. Multi-view (multi-scopic) visualization or free-viewpoint type applications necessitate the virtual view synthesis. View synthesis is performed, since the input source (sparse multi-camera views) is not enough to generate a continuous parallax when projected on the display. Different approaches have been adopted by different visualization systems, some of which carry out virtual view synthesis solely based on the texture information (e.g. camera views) and some of which perform view synthesis with the aid of scene depth information (simply referred to as 'depth') in addition to the texture information. Since the source camera views are transformed in the coordinate system of the virtual camera view, in the rendering process of complex scenes (e.g. with multiple video objects at relative locations with respect to each other and with respect to the camera), disocclusions may appear that need to be in-painted properly to overcome any perceptual degradations.

Unlike offline (non-real-time) processes, such as movie production, virtual view synthesis can also be integrated within the live and/or real-time 3D broadcasting and streaming chains featuring modern 3D display systems, the details of which are depicted in Section 5.4. In the latter case, it is essential that the video rendering engine performing 3D virtual view synthesis, that is not limited to the synthesis of a single virtual camera, operates in real time. Hence, particular hardware accelerated GPU architectures need to be deployed in carrying out complex real-time multi-view rendering or free-viewpoint video rendering tasks, which is a trendy research topic since the uptake of 3D displays industry as well as the successful progress in 3D multi-view acquisition, compression and transportation research.

5.2 3D Video Rendering

3D video rendering can be implemented on a STB (set-top box), or as any other software processing block to run (if compliant) in a GPU. A 3D video

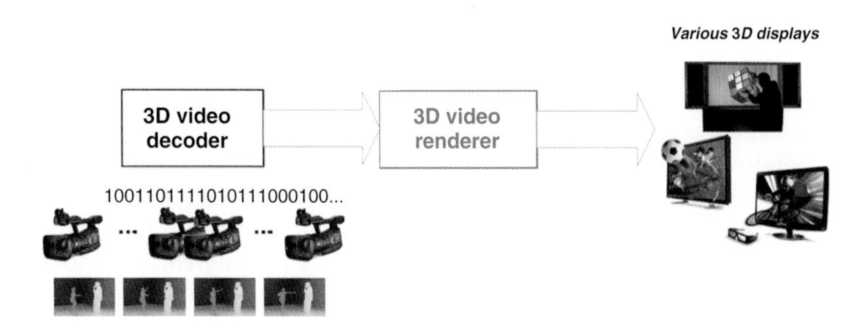

Figure 5.1 3D video rendering input and output

rendering engine has its input data channels delivering raw video from the 3D video decoder, such as an MVC decoder, or multiple 2D video decoders, each delivering a different decoded camera view, and provides output signals to various 3D display technologies that will be outlined in Section 5.4. Figure 5.1 depicts this relationship.

In this section, depth-aided view interpolation complemented with occlusion handling, in-painting, de-aliasing, and displaying specific view scrambling will be discussed. Note that the multi-view video rendering does not necessarily have to be aided with explicit scene depth and geometry information. There are a number of works in the literature, which outline how image-based multi-view rendering works without any prior geometry knowledge, or with implicit and limited geometry information. The authors' survey in [1] outlines several conducted works on image-based rendering in detail by explaining the underlying key technical frameworks.

Image-based rendering techniques are categorized based on how much geometric information are used in them, i.e. whether the method has used explicit geometry (e.g. per-pixel disparity/depth, or layered depth information), implicit geometry or stereoscopic correspondences, or no geometry at all. On the other hand, the 3D viewing experience is not solely dependent on the accuracy of the use of geometric constraints during rendering, but on a combination of the technical, perceptual and artistic constraints.

Most recently, the significant perceptual aspects of stereo vision and their implications for stereoscopic content creation have been studied in detail to derive perceptual-aware disparity mapping operators, which are locally adaptive, based on saliency in the video [2]. The work presented in [2] computes disparity and image-based saliency estimates from a sparse set of stereo-correspondences, which are then used to deform input views in a non-linear fashion so as to meet the target disparities. No explicit knowledge of camera calibration parameters, per-pixel dense depth maps, occlusion handling, or in-painting is necessary as opposed to most depth-based 3D rendering schemes.

On the other hand, there are several advantages of exploiting dense disparity/depth map information. One of them is that it is possible to synthesize virtual target views at a further distance from the source (real) camera views with sufficient perceptual quality thanks to the disparity information associated with each pixel. Hence, multi-view rendering at a wider baseline is possible, resulting in enhanced movement parallax. Second, the use of decoded dense disparity/depth information on the client terminal side avoids the time-consuming computation of disparity correspondences from decoded textures of multiple views in order to perform rendering. In the rest of this chapter, the depth-based view interpolation is explained.

Several different 3D video use cases can be listed depending on the source multi-view format, delivered service and the 3D display technology used, though two scenarios will be considered in this chapter. In the first scenario, the user is delivered a stereoscopic video (i.e. the left view video and the right view video) and the associated per-pixel depth maps of the delivered stereoscopic video. The user is free to adjust the depth on the stereoscopic display, so that the disparity between the viewed left and the right pairs can be changed. The left and the right views are used in combination with the depth maps to create an interpolated view in between, where the positioning of this interpolated view determines the depth level in the watched stereoscopic content.

In the second scenario, the user is assumed to have a multi-view display that displays multiple viewing angles simultaneously. More information on the 3D display technologies is provided later in this chapter. In the second scenario, the user is delivered two more camera streams in addition to the stereoscopic video-plus-depth pair. These cameras are located on either side of the stereoscopic camera pair with varying baseline distance from the stereoscopic camera pair. It is assumed that all cameras are vertically aligned and that the disparity is only horizontal. Figure 5.2 depicts the camera configuration.

The stereoscopic baseline is adjustable in both scenarios. Typical boundaries are 3 cm (lowest) and 7 cm (highest), where the adjustment is typically done with the consideration of stereoscopic vision cues. The inter-axial

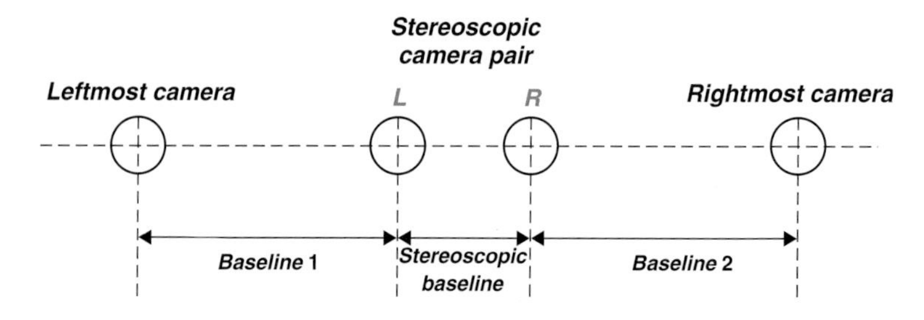

Figure 5.2 Stereoscopic 3D video rendering input and output

distance that specifies the stereoscopic baseline controls the overall range of depth, or in other words, the depth volume of the reproduced scene. In order to create proper stereoscopic video, a variety of rules need to be considered, one of which is to ensure that the captured scene is fitted within a Comfortable Viewing Range (CVR) of the viewing context, such as the ratio of screen width and viewing distance [2].

To facilitate a comfortable long-term stereoscopic viewing experience, all objects in the scene need to stay within this limited depth space close to the screen. Baseline 1 and Baseline 2 depicted in Figure 5.2 represent the distance of the leftmost camera and the rightmost camera from the stereoscopic camera pair in the centre. The outer cameras may or may not be equidistant from the centre camera pair and usually the Baseline 1 and Baseline 2 are larger compared to the stereoscopic baseline. The larger those baseline values, the wider the 3D reconstruction space and increased horizontal parallax. The view interpolation processing required to address a multi-view display depends on the display itself, such as the total number of views that the specific display supports. The distance between each view-pair on the 3D display can be calculated in two different ways. In the first assumption, the stereoscopic baseline is regarded as the reference. Then, depending on the number of views supported by the target 3D display, the positions of the views to be synthesized within the total camera baseline are determined. Figure 5.3 shows the respective calculation for a 5-view display and a 9-view display.

In the second assumption, it is aimed at conserving the complete baseline of the multi-camera set-up, which is the total of Baseline 1, Baseline 2 and the stereoscopic baseline depicted in Figure 5.2. The distance between each view-pair depends on the complete baseline of the multi-camera set-up as well as the total number of the views that the 3D display supports. Figure 5.4 depicts the calculation for a 5-view and a 9-view 3D display in the second case.

Virtual views within the total camera baseline are interpolated using two camera views and the corresponding depth maps, which are first converted to disparity values. For a target virtual view, one reference is on the left-hand

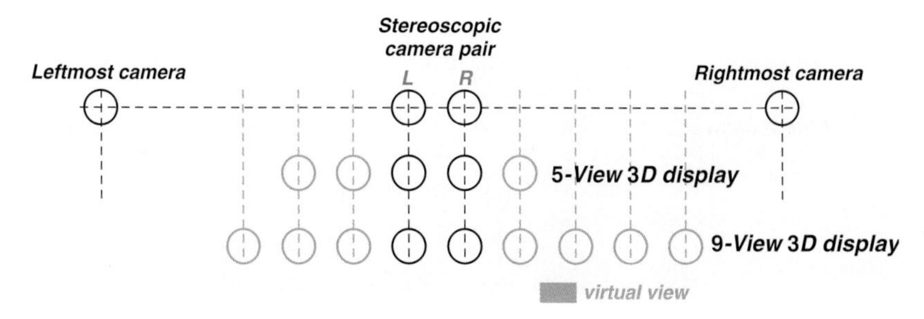

Figure 5.3 View interpolation taking the stereoscopic baseline as the reference

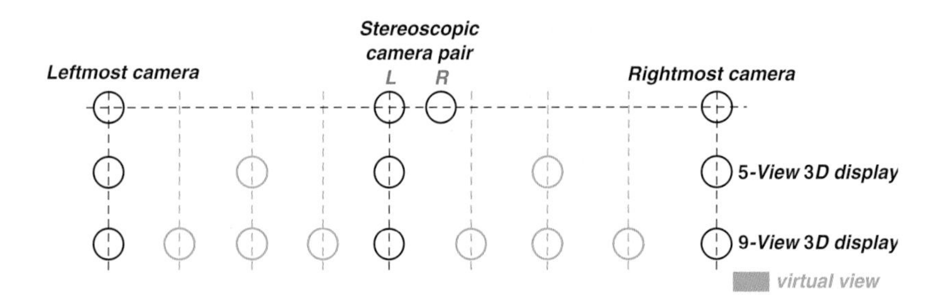

Figure 5.4 View interpolation exploiting the complete multi-camera baseline

side of the virtual view and the other reference is on the right-hand side of it. First, the left and the right reference images are segmented into a main layer, a foreground layer and a background layer based on the edge characteristics of the depth maps and the texture of both reference views. The major segment is the main layer that comprises all pixels that are not located in the vicinity of edges. The foreground layer comprises the pixels that are in the close vicinity of the edges and whose depth value is lower than that of the pixels on the other side of the edge, i.e. closer to the camera. The remaining pixels are classified as the background layer. Figure 5.5 depicts an exemplary segmented frame from the multi-view shooting of one of the European Union (EU) FP7 projects MUSCADE (Multimedia Scalable 3D for Europe) [3]. Grey pixels represent the main layer, white pixels represent the foreground layer and the black pixels represent the background layer.

The main idea behind segmenting the reference views first is to differentiate between reliable and unreliable depth values that are located across the boundaries of sharp edges (i.e. depth transition areas). Subsequently, the unreliable regions that consist of the foreground and the background layers are treated separately from the main layer during the forward warping and blending processes. Canny edge detector can be used to mark an area as unreliable around the detected edges, where finding the threshold value for the edge detector depends on the shot multi-view content and

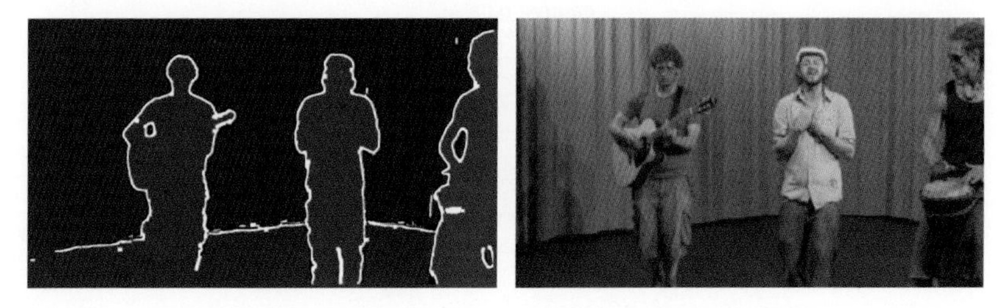

Figure 5.5 Segmented image on the left side and the texture image on the right side

can be determined offline to be counted as a metadata to be used by the video renderer.

The differentiation between these three distinct areas basically leads to the removal of artefacts propagated by the background layer while forward warping (i.e. interpolation). This layer contains texture pixels from the foreground wrongly associated with background depth values. Similarly, texture pixels from the background that are wrongly associated with foreground depth values also exist, which, however, do not lead to visual distortions, since they are correctly replaced with the correct foreground pixels. Each extracted layer is projected onto the interpolated views coordinates. For interpolation, first, the depth values are translated into disparity that represents the distance the corresponding pixel needs to be moved to be projected onto a target view.

$$d = \frac{f \cdot l}{z} + du \tag{5.1}$$

In Equation (5.1), d represents the disparity, whereas f stands for the focal length of the cameras (the focal lengths of all cameras in the multi-camera set are assumed to be identical), l represents the baseline distance between the reference camera position and the position of the view to be interpolated, z denotes depth and du stands for the difference in principle point offset. Figure 5.6 illustrates the projection of the disparity values of the reference left-view layer (either main layer, foreground layer, or background layer) onto the grid of the interpolated view. For the ease of illustration, only a 1D grid is considered, since there is disparity only in the horizontal direction.

During the disparity projection, information on the occluded areas (part of the 3D scene that is not visible in one of the views) is determined. Figure 5.7 outlines the occlusion cases. Then, the texture information in the left reference and the right reference views that is pointed out by the interpolated disparity values are used and blended together, depending on the occluding area information to render the interpolated view. While the disparity values of the left reference view are projected, each disparity value is first scaled. Note that the original disparity values have been calculated taking the distance

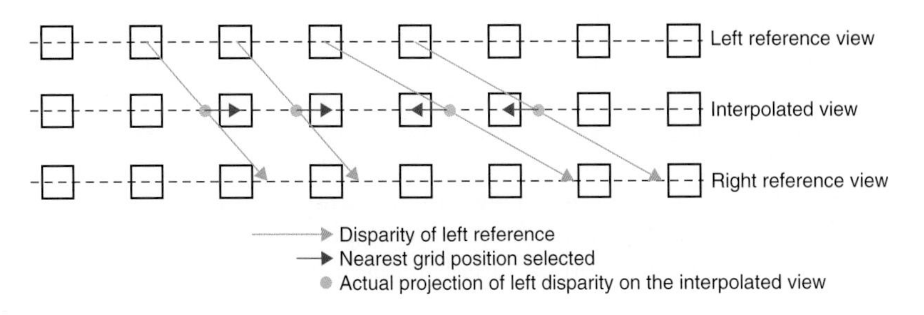

Figure 5.6 Projection of the left reference on the interpolated view

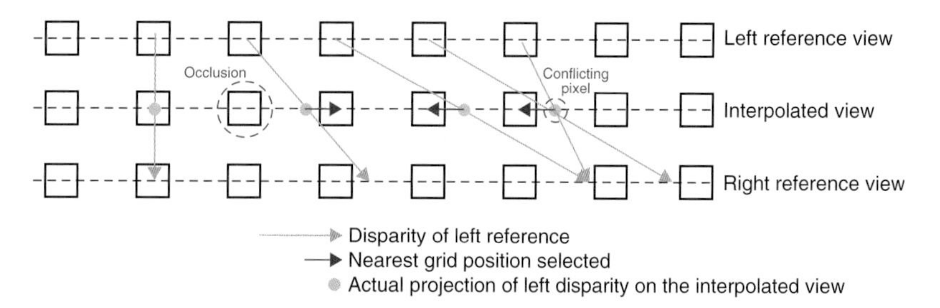

Figure 5.7 Illustration of occlusion and conflicting pixels during disparity projection

between the left and the right reference views as the baseline parameter in Equation (5.1). The original disparity values are multiplied with a coefficient in the $[0, 1]$ range, in relation to the proportion of the distances of the left and the right reference views from the interpolated view. Then, each disparity value is projected at the position defined by the scaled disparity. This position is rounded to the nearest pixel location as shown in Figure 5.6.

In order to remedy conflicting pixels shown in Figure 5.7, where different disparity values point to the same pixel position, the disparity values of the left reference view are scanned from left to right so that information corresponding to the occluding pixels overwrite the information corresponding to the occluded pixels. In order to fill the disoccluded areas (i.e. the created holes), each projected disparity value is automatically assigned to all unfilled pixel locations on the right-hand side of the previously filled pixel position up to the pixel position that the projected disparity value points at.

It is necessary to identify the regions of the interpolated view that are only visible in one view in order be able to render the texture for them using only this reference view, because they are occluded in the other view. Information on what is occluded in the right reference view is obtained by comparing the disparity values in the left reference views disparity map with the disparity values they point to in the right reference views disparity map. The difference between the two indicates visible objects in the left reference view, but which are occluded in the right view, or vice versa. Similarly, information that is occluded in the left reference view is obtained by the same process, comparing the disparity values in the right disparity map and the disparity values they point to in the left disparity map.

The identified occlusion information tells us which view (left reference, right reference, or both, or none) can be used to render the texture of each pixel in the interpolated view. In the interpolation process, unidirectional interpolation is used for occluded pixels, whereas bidirectional interpolation is used for non-occluded image areas. However, such a process results in some visible artefacts in the interpolated view image in the depth transition areas, such as the sudden transition between the pixels filled

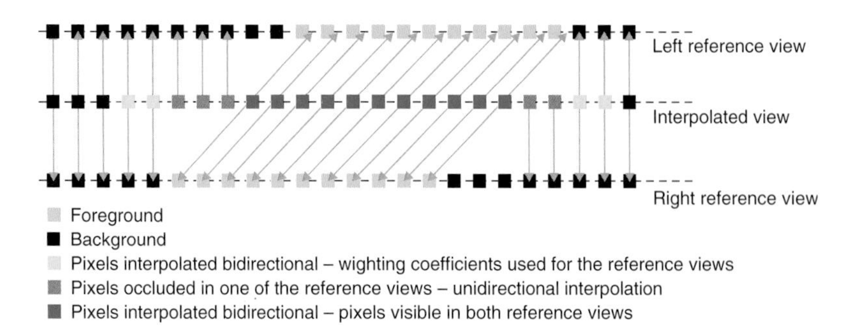

Figure 5.8 Interpolation scheme taking into consideration the occlusions

with unidirectional interpolation and the pixels filled with bidirectional interpolation. The disparity map may not perfectly match the borders of video objects.

Most of the time, it is difficult to track the transition from a background region to a foreground region in a sharp way due to the slightly blurred edges, which therefore leads to kind of a disturbing ghosting effect in the interpolated view images unless some caution is taken. In one of the recent European projects, called ROMEO (Remote Collaborative Real-Time Multimedia Experience over Future Internet) [4], the transitions between the unidirectional and the bidirectional interpolation are performed in a weighted manner depending on the pixels' location as a cautionary step during the interpolation process. In other words, the coefficients used for bidirectional interpolation are adapted based on the position of the occluded pixels. The coefficient management depends both on the width of the occlusion space and the depth difference between the background and the foreground layers. Figure 5.8 illustrates the interpolation process taking into account occlusion handling and bidirectional interpolation with weighting factors.

Another visual degradation that results after the view interpolation is the aliasing effect, which is visible on the borders of the synthesized objects. The reason for this is the association of a single disparity value to each of the interpolated pixels. Thus, the borders of the synthesized objects in the interpolated disparity map sharply coincide with the pixel borders and so do the object borders in the texture of the interpolated view. Aliasing should be efficiently overcome by means of post-processing. In the first step, the borders need to be detected in the interpolated disparity map and subsequently a sub-pixel mask needs to be adapted, based on the content in order to filter the contours. As a result, each sequence of pixels along the object borders is associated with a couple of disparity values, like semi-transparent objects. Figure 5.9 shows an example of a rendered texture, before and after de-aliasing.

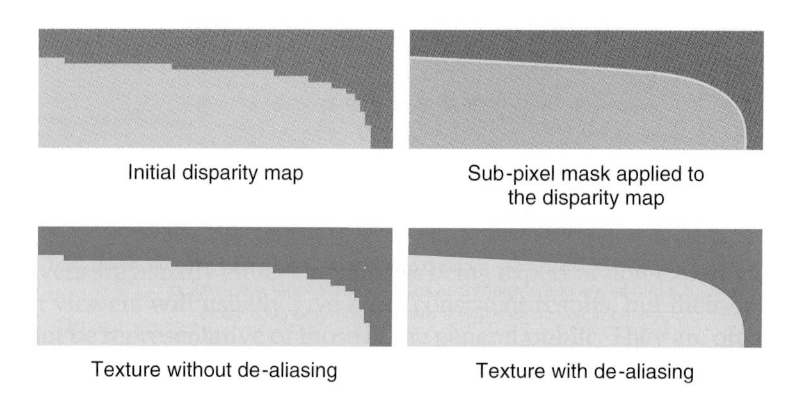

Figure 5.9 Interpolated texture samples with and without de-aliasing (5)

In the final stage of 3D video displaying, scrambling needs to be applied on the interpolated and rendered texture images, depending on the particular display type. For stereoscopic 3D video content, specific processing is applied instead to ensure interoperability, i.e. to convert the rendered content into a universal format accepted by most standard 3D displays. Side-by-Side or Top-Bottom formats are among the most common display formats for stereoscopic 3D video content, which was previously mentioned in Chapter 1. These are also known as "frame-compatible" formats, where the pair of images belonging to the views from the left eye and the right eye are downsampled in space by two in one of the x or y directions, and then combined to produce a single image that has the same spatial resolution as the original left eye and right eye images. Another format that is known as "frame sequential" is also available, which does not involve any sub-sampling of images, but consists of a sequence of alternating frames wherein each successive frame carries the image meant for the right or the left eye. Figure 5.10 depicts the formation of these formats considering the original stereoscopic image pairs at Full-HD resolution (1920 $\times$ 1080).

To address multi-view displays, the details of which are discussed in Section 5.4, the display scrambler is usually directly linked to the display's specific pixel arrangement. A multi-view display can show a different number of views and thus a different scrambling operation is necessary depending on the display type. Moreover, multi-view displays show reduced resolution images to the different directions. Due to the way they are constructed, they allocate a fixed amount of pixels available in the underlying display panel to certain viewing directions. For instance, if a multi-view display can show nine views, then the number of pixels shown in one direction is approximately one-ninth of the full display resolution. The loss in the spatial resolution is usually evenly distributed between the horizontal and vertical resolutions. Figure 5.11 outlines the overview of the 3D rendering framework for multi-view displays, including the sub-sampling and tiling.

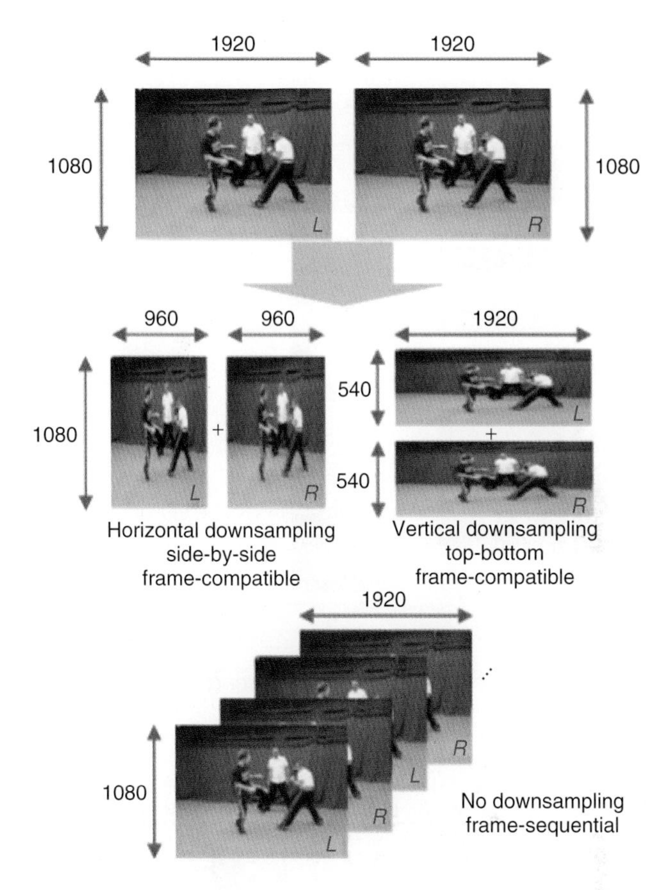

Figure 5.10 3D stereoscopic video display formats (5)

Most display manufacturers prefer to keep the specific pixel/sub-pixel structure of the shown multiple views a secret, and users therefore are forced to use proprietary tools in order to be able to display 3D content. For such displays, developers and researchers are commonly forced to reverse engineer the sub-pixel structure [6]. The straightforward way of representing display-specific sub-pixel patterns is to create a texture map, which has the same size as the underlying display panel, where the view numbers are represented by R, G, and B. The generated texture in the renderer is used as a look-up table during the assembly of the final multi-view image (i.e. scrambling). Figure 5.12 shows a sample display texture map for the Newsight multi-view display that is capable of displaying eight views. According to the representation in Figure 5.12, the pixel at location $x = 1$, $y = 1$ (top-left corner) is sampled using views 1, 2, and 3. The red sub-pixel is sampled from view 1, whereas the green and blue sub-pixels

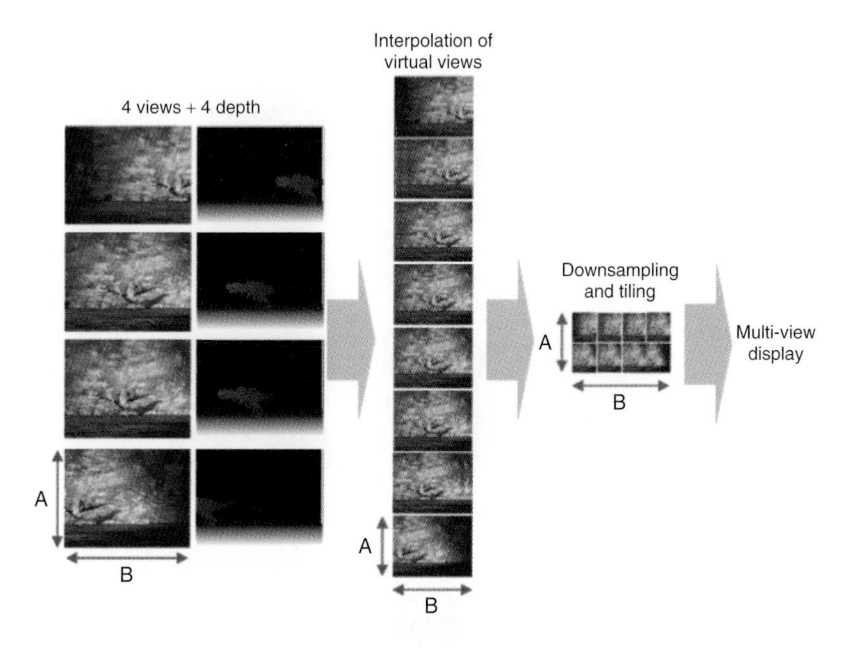

Figure 5.11 Overview of 3D video rendering for an 8-view display

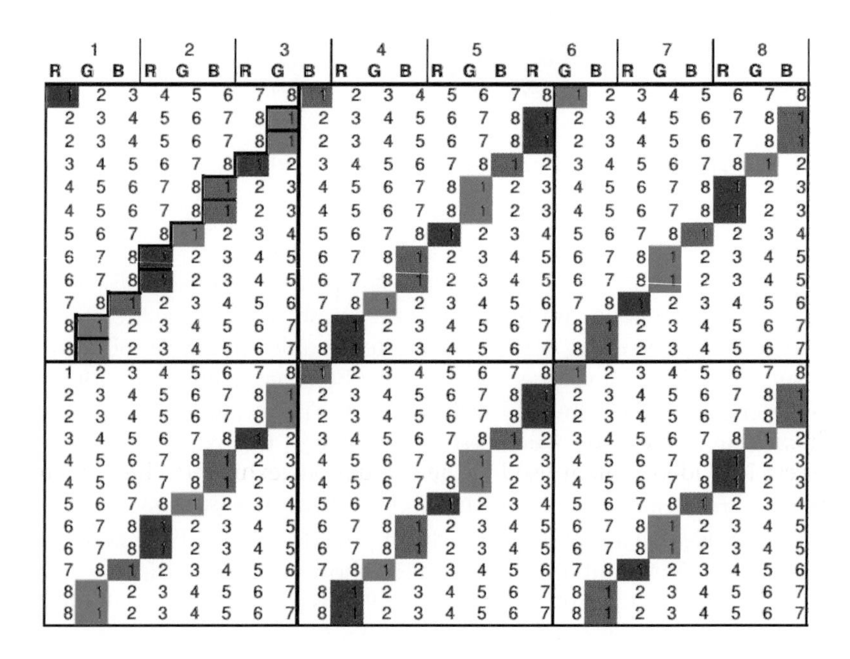

Figure 5.12 Pixel order structure for the Newsight 8-view display (7)

are sampled from views 2 and 3, respectively. The resolution of the specific display under concern is 1280 × 1024.

5.3 3D Video Adaptation

3D video adaptation requires in-depth investigations in order to enable efficient 3D media delivery systems that satisfy the needs of media consumers in a highly heterogeneous environment of access technologies and 3D display devices. Furthermore, compared to 2D media adaptation technologies, 3D adaptation schemes also need to take into consideration the greater availability of different representation formats and application requirements.

The overall target of the 3D media adaptation is to maximize the users' experience in terms of perceptual quality (image quality and depth perception together), which is affected by several factors, including 3D content characteristics, contexts prevailing in various use environments, etc. In the rest of this section, first, the importance of depth in 3D video adaptation and the context-based adaptation are explained. An overview of 3D video adaptation facilitation for mobile networks is presented afterwards. Finally, the key elements of network adaptation in multi-view 3D video delivery systems are outlined.

5.3.1 Importance of the Depth Map in Adaptation

As discussed in Section 5.2, the depth map information is essential in 3D visualization. Thus, unlike in generic 2D media adaptation, the 3D video adaptation process needs to consider depth map information. The impacts of the spatial resolution of depth maps, which were encoded at different quality levels, on the perceived quality and depth perception of 3D video have previously been evaluated in [8]. In this work, owing to its flexibility and compatibility with the existing coding and transmission technologies [9], colour-plus-depth (also known as video-plus-depth) based 3D video was used as the representation format to perform the evaluations.

Scalable Video Coding (SVC) [9] has been used to code the depth maps. The view (texture) counterpart of the depth maps has been used without encoding in order to assess the scaled depth maps, so as to avoid any potential influences from the colour images on the overall 3D video quality evaluation. The test 3D video content used in the evaluations exhibited different structural and motion complexities. The evaluation results have demonstrated that the higher the spatial resolution, the higher the perceived video quality and depth perception are. When the structural complexity of a depth map increases, the perceived quality saturates while the depth perception is enhanced at higher qualities. Furthermore, when the motion complexity in the 3D video sequences increases, the effects of higher spatial resolutions become more significant on the video quality and depth perception.

5.3.2 Context Adaptation

In the work presented in [10], the effects of varying ambient illumination context in the content usage environment on the viewers' 3D quality of experience were investigated with a group of viewers. The results of these investigations had revealed that ambient illumination has significant effects on the perceptual video quality and depth perception of 3D video. It had been reported that when the ambient illumination in the usage environment increases, the Mean Opinion Score (MOS) ratings of the viewers also increase. On the other hand, when the ambient illumination increases, the depth perception ratings of the viewer decrease. These observations had provided a notable set of findings to develop 3D video adaptation techniques for the 3D video delivery systems on the internet, including Content Delivery Networks (CDN) and P2P video streaming systems, as well as to mobile 3D consumers, for whom the ambient illumination conditions can change dynamically while the user is active. Such a 3D video adaptation technique was outlined in [11]. In the presented work, a user perception model was proposed to characterize the video perception of users towards 3D video contents examined under a particular ambient illumination condition. Motion, structural complexity, luminance contrast, and depth intensity characteristics of various 3D test videos were regarded as the main content-related contextual information in the presented model. Using this model, the adaptation technique used to determine the bit-rates of the target 3D video contents under changing ambient illumination conditions was proposed.

5.3.3 3D Video Adaptation for Mobile Terminals

Parallel to the uptake of mobile computing technologies and the tremendous increase in the rate of video consumption through mobile internet, technologies enabling 3D viewing on a mobile, as well as compressing and delivering 3D video sequences to mobile users efficiently over a variety of mobile access networks have attracted a lot of attention lately. On the other hand, the 3D video experience similar to its predecessor, 2D media, is more susceptible to disruptions, cuts and other sorts of artefacts on mobile conditions than in fixed networks. Thus, deploying perceptual-aware and efficient dynamic 3D media adaptation mechanisms especially on the nodes connecting the mobile terminals to the internet (such as mobile base stations) is essential for a smoother experience for mobile 3D video consumers.

The Media Aware Proxy (MAP) is a transparent user-space block that is deployed for low delay filtering of 3D video streams delivered to mobile users, as shown in Figure 5.13. The need for such a middleware in the base stations serving mobile 3D customers, or close to base stations in the network, derives from the fact that each mobile user will be able to seamlessly hand over between different access networks instantaneously available to them (e.g. 3G LTE, Wi-Fi) without a reduction in the perceived

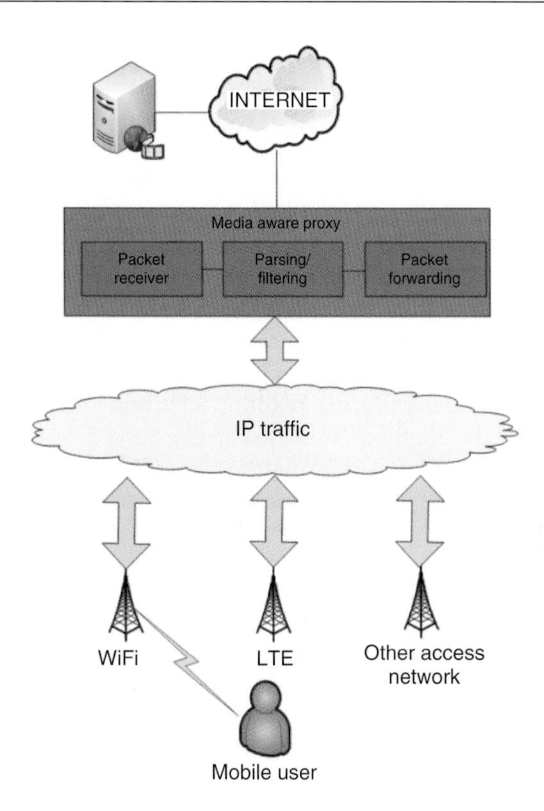

Figure 5.13 Mobile media traffic through the media aware proxy (14)

3D Quality of Experience (QoE, explained in detail in Chapter 6) as a result of changing network capabilities. Therefore, in cases of a handover to a new access network, MAP either drops or forwards packets that carry specific sub-streams of a scalable 3D stereoscopic bit-stream to the receiving mobile user. Its functionality extends from the network layer to the application layer providing video stream adaptation by taking into account the mobile clients' network conditions. Delivered packets through the internet are forwarded to their internal routing module, where the packets' header fields are parsed without altering them to identify content-specific information (application level). The packet discarding mechanism, or the adaptation decision-making mechanism, is a fundamental part of MAP.

This functionality periodically decides which sub-layers to drop from the forward stream, based on the currently available bandwidth to the mobile client. In order to augment the operation of such a middleware tool, it is necessary to deploy scalable 3D stereoscopic video coding (e.g. SVC [9]) that comprises several different quality layers for both the left and the right eye channels. The same concept has been outlined in [12]. A network distribution model that takes advantage of the multi-layered characteristics of scalable video is the multicast with a Media-Aware Network Element (MANE) that aggregates or trims multiple RTP. Each RTP session carries sub-stereoscopic video information, such as different quality layers of either the left or the right eye view. MANE generally resides between the media server and the clients.

The 3D media server performs transmission over multiple RTP sessions. MANE is in charge of collecting the RTP sessions, de-packetizing them, customizing Network Abstraction Layer (NAL) units according to the mobile consumers' needs through adaptation, and subsequently delivering to them a single RTP session over a single transport address. MANE can get 3D content-related important information for adaptation decision-making using either signalling, or generally via reading RTP and NAL unit headers on the fly. Especially the methods depicted in Sections 5.3.1 (depth-based 3D videos) and 5.3.2 (contextual information-based 3D video adaptation) can be deployed in the adaptation decision-making process in the MANE block either in a feed-back based or feed-forward adaptation manner.

5.3.4 Multi-View Video Adaptation

Multi-view video streaming over the internet has gained attention with the advances in multi-view capturing, compression and visualization technologies. As mentioned in Chapter 2, Multi-View plus Depth (MVD) is a popular format that enables the rendering of multiple arbitrary camera views from a limited number of source camera views accompanied by the associated per-pixel dense depth information. However, the delivery of MVD data over the existing aerial broadcasting infrastructure (e.g. DVB-T2, DVB-S2) is not feasible for a number of reasons.

The first reason is the theoretical capacity limitation associated with the access technology. Not a high enough number of multi-view video channels especially comprising more than two camera views with the associated depth, occlusion, transparency information, etc., can be accommodated in the network. The current trend in Digital Video Broadcasting (DVB) for 3DTV services indicates that it is not planned to standardize multi-view services comprising more than two camera views with depth information in the foreseeable future. In the first generation of 3DTV services over DVB, stereoscopic 3D video signals comprising the left and the right eye views are transmitted either in a Frame-Compatible (FC), or in a Service-Compatible (SC) manner. FC refers to converting the stereoscopic 3D video into a single HD video stream, whereas SC refers to transmitting the extra view as a separate stream, probably encoded with a non-backwards compatible standard (e.g. MVC), where the base view is encoded in a backwards compatible way.

The second reason why it is not feasible to broadcast multi-view video content with depth over DVB is the heterogeneity in multi-view consumption environments and applications. Not every type of 3D display would necessitate the presence of the same amount of source (input) camera views to create good quality 3D video, which will be discussed in Section 5.4. Hence, it is conceptually inefficient to allocate spectral resources for a certain amount of video information that may be useful to only a group of users. This fact emphasizes the importance of delivering multi-view services to users in a customized way, for which delivery over the internet is the best mechanism. Nevertheless, although the internet allows the users to access the 3D multi-view content using flexible transport mechanisms and protocols, there are a number of factors that prevent the continuity of the service at maintained levels of 3D QoE, unless certain precautions are taken. The primary factor is the inherent network congestion due to the background internet traffic (e.g. other services) and the corresponding data packet losses. Regardless of the underlying delivery mechanism of multi-view video streams (e.g. P2P delivery, delivery over CDNs, others), deployment of perceptual-aware network adaptation mechanisms is necessary in order to dynamically adjust the streaming rate between the two nodes (e.g. peer or router) on the delivery path.

In a multi-view video delivery scenario, where the content is encoded so that the multi-view bit-stream exhibits view scalability (such as simulcast coding as explained in Chapter 3) and quality scalability (e.g. SNR scalability in SVC), it is possible to achieve a useful margin of operating points depending on the instantaneous network state. On the other hand, in order to select the best operating point against the changing bandwidth conditions, a number of factors need to be taken into consideration that leads to an adaptation that is aware of both the content and the application. These are:

1. *Display type* – It might be unnecessary to deliver more than two video-plus-depth pairs to a client with a standard 3D display capable of displaying a stereoscopic view at one time only. On the other hand, the demand of an 8-view auto-stereoscopic multi-view display and a holographic display, which can display more than a hundred viewpoints, on the number of input camera views would be different.
2. *Users' preferred viewpoint in the free-viewpoint viewing context* – Even though a client may have a standard 3DTV, within the free-viewpoint viewing context, they would have the opportunity of changing their viewing angle. This mode of application would necessitate updating the list of viewpoint streams actively delivered to the client. For instance, camera views that are closer to the viewpoint of interest would take higher chance of reconstructing the 3D video with a higher fidelity, and thus would take priority in the delivery system.
3. *The perceptual influence of different quality layers of the compressed viewpoints on the virtual view of interest* – If the network conditions are severe, which would not allow the delivery of two camera viewpoints of the highest possible quality, it is important to select the group of quality layers to drop from different video-plus-depth pairs, so that the perceptual quality of the

synthesized stereoscopic viewpoint stays at the highest possible quality. This information can be computed at the server side that potentially has access to the originals of the camera views and hence can exhaustively measure the perceptual influence of each generated quality layer on a number of target virtual viewpoints. Subsequently, this information can be embedded in the multi-view transport stream to be exploited in the relevant network nodes when adapting the multi-view stream for the group of clients.

In addition to the factors described above, in the context of multi-view and/or light-field displays, the network conditions might sometimes necessitate dropping the streams of several video-plus-depth pairs overall to match the network rate. In such a scenario, the key to the adaptation decision-making process is to select the best set of camera views, so that the average distortion of all delivered views and the viewpoints that have to be synthesized or concealed from the viewpoints that have been delivered is minimized. Similar to the third factor explained before, it might be possible to compile a set of adaptation hierarchy information on the server side that would be delivered alongside the multi-view video stream in order to facilitate quality-aware adaptation through the delivery path.

5.4 3D Display Technologies

Common to all modern 3D display technologies is their capability to bring the viewers the perception of stereoscopic 3D depth. This is basically achieved by exposing to the left and the right eye a pair of offset two-dimensional images, where there is a slight disparity between both images. The perception of depth is a result of the visual processing of the pair of images in the human brain. On the other hand, beyond the visualization of stereoscopic depth is the perception of additional spatial information related to the observed scene. This involves the sensation of surroundings and occluded objects. However, this is not necessarily addressed by many of the existing 3D displays, which display the scene from a particular viewing angle only, regardless of the relative position of the head with respect to the screen.

Stereoscopic displays, which usually necessitate wearing special eye glasses, are of that type. Three of the most commonly known types of stereoscopic displays (one of them being just a type of stereoscopy, but not specifying a particular display type) are depicted accordingly in Section 5.4.1 and Section 5.4.3. Some other types of displays, which do not require any special eye glasses to be work in order to watch, can also improve the spatial parallax with respect to stereoscopic displays by interpolating and extrapolating additional viewpoints and exposing them from the display along different angles. The auto-stereoscopic display technology is depicted in Section 5.4.4, whereas the light-field displays are explained in Section 5.4.5.

5.4.1 Anaglyphic Stereoscopic Displays

Anaglyph stereo does not specify a particular display technology, while any conventional 2D display can display anaglyphic stereoscopic image/video. The audience does not need 3D-compatible TV or projector equipment. Regardless of the display technology, as well as regardless of the frame rate, aspect ratio, and spatial resolution, anaglyphic stereoscopy can be experienced. If you can watch a movie, you can watch an anaglyph movie in 3D.

Anaglyph stereoscopy refers to the stereoscopic visualization effect (i.e. perceived depth as previously described) achieved by letting the left and the right eyes see the offset stereoscopic video pair using special colour filters. The colours must be chromatically opposite. Red and cyan are a typical example of anaglyph colour filters. When someone observes the superimposed and colour-coded video pair behind the so-called "anaglyph glasses", an integrated stereoscopic view is revealed. The brain's visual cortex then fuses this into the perception of a three-dimensional scene.

Anaglyph stereoscopic viewing is so far the cheapest way creating the 3D depth effect, either on a printed magazine or on a display/monitor. Coloured glasses (see examples in Figure 5.14) are cheap to produce. In addition to that, it is possible to display a stereoscopic video in full 3D resolution on a 2D screen, simultaneously for both eyes.

However, there are several disadvantages of anaglyph stereoscopic that hinder the perceived depth significantly. For instance, if some colour information from the left eye video gets mixed into the right eye video (and vice versa), a slightly coloured ghost is seen, where the increase in the parallax (i.e. the disparity between the left and the right eye views) separates the seen stereoscopic video pair more and makes ghosting worse. Furthermore, if there is a brightness level between the two pair of views, the resulting flicker-type effect is rather disturbing. The eye that sees the darker view feels dull. Elderly people have difficulty in changing the focus of their eyes. Consequently, different colours may come to focus at different distances, disturbing the stereoscopic sight. Because of the conflict between the focus levels of both eyes, this also results in eyestrain and triggers headaches. Some anaglyph glasses correct dioptre to cure this problem. Last but not the least, a major disadvantage of anaglyph stereoscopy is the fact that it is

Figure 5.14 Different styles of anaglyphic glasses (most commonly used are the red-cyan pairs that are on the left and in the middle, green-magenta pair on the right)

impossible to get the colour information exactly right, because each eye is delivered only part of the Red-Green-Blue (RGB) colour range. The viewer is always subject to a kind of partial colour blindness and because of its poor colour definition the application areas of anaglyph stereoscopic have been limited. Anaglyphic stereoscopy is more successful in displaying less colour-rich images and videos, in black and white for instance.

5.4.2 Passive Stereoscopic Displays

This group of displays depicts a particular display technology unlike the anaglyphic stereoscopy presented in the previous section that is realizable with conventional 2D display sets and projectors. In passive stereoscopic displays, special films are applied to the display screen to produce the 3D depth effect, and a pair of "polarized" glasses separates the images that are emitted from the screen for both eyes. It is a passive technology, as the glasses do not require any circuitry driven mechanism to handle the polarization of the stereoscopic video being shown. In that sense, it is very similar to the colour-filter glasses used for anaglyphic stereoscopy. Furthermore, it is the technology that is typically used by cinemas showing modern 3D movies (e.g. RealD 3D equipped motion picture theatres). For instance, the popular RealD 3D technology uses circularly polarized light beams to reveal stereoscopic video projection. The advantage of circular polarization technology over linear polarization is that viewers are able to tilt their head without sensing the 3D scene as distorted (e.g. double images).

In the current 3DTV market, passive stereoscopic displays can be regarded as the most common solution for delivering the stereoscopic 3D video experience to the audience. As opposed to the projection system, displaying a stereoscopic view with different polarizations on a flat-screen home display is a bit more complicated and necessitates coating the screen with a special film. This film polarizes the screen image and splits what your right and left eye see when combined with the polarized glasses. It ensures that the correct eye sees the corresponding video. Usually, the displays that support passive stereoscopic 3D viewing do not provide higher display rates than other conventional displays and TV sets (i.e. 60 Hz). The displayed picture comprises the left eye and the right eye views simultaneously, each of them being at half resolution. In other words, the passive polarization technology delivers half of the horizontal lines (vertical resolution) per eye, meaning 1920×540 pixels per eye. This fact has been cited as the most notable limitation of passive stereoscopic 3D methods, as if this limitation results in a lack of support for Full-HD video content.

The Film-type Patterned Retarder (FPR) technology, patented by LG Display and known to be a leading technology in the market, has adopted "space-division" and displays both the left and right eye image at the same time by using lines of resolution. This fact is sometimes negatively interpreted such that passive polarized stereoscopic 3D displays cannot display Full-HD (1920 × 1080) resolution content. However, a viewer is still able to see a Full-HD video, when the left-eye view is combined with the right-eye view. Another drawback with the special film-coated passive stereoscopic 3D displays is that the black bars that separate the horizontal and vertical stripes on the passive filter can sometimes give rise to notable and visually distracting stepping effects on the displayed stereoscopic view.

5.4.3 Active Stereoscopic Displays

Unlike the passive stereoscopic 3D displays that are presented in the previous section, in active stereoscopic 3D displays, it is the special liquid crystal "shutter" glasses that have the primary role in creating the 3D depth effect. Furthermore, this kind of displays deploys a "time-division" stereoscopic 3D displaying mechanism unlike the "space-division" method used by passive stereoscopic 3D displays.

Two other major differences from passive polarized stereoscopic 3D displays are: the displays need much higher temporal resolution (i.e. refresh/frame rate) in order to be able to divide the temporal frequency for both eyes and the eye-glasses require special material and hardware circuitry. In Figure 5.15, a pair of a passive-circularly polarized glasses and a pair of active shutter glasses are shown side by side. It can easily be seen that active shutter glasses look heavier and more complex than passive-polarized glasses. Active stereoscopic displays arrived on the 3D consumer display market earlier than their passive stereoscopic competitors.

The glass for both eyes contains a liquid crystal layer. This special layer has the ability to change from light to dark (and vice versa) when electric voltage is applied to it. Exploiting this ability of the liquid crystal glasses, both eyes' glasses are periodically lightened and darkened in an alternating manner, and in synchronicity with the 3D displays' screen refresh cycle. The 3D display alternately displays the left-eye and the right-eye views in a time-divided manner at a higher frame rate than the original stereoscopic video.

The synchronization is typically accomplished using a stereo connector attached to the graphic processor that communicates either directly with the glasses or via a wireless. This communication comprises sending the appropriate timing signals for switching the glasses on and off. The glasses

Figure 5.15 RealD 3D style circularly polarized glasses (left), and generic active shutter glasses (right)

switch on and off at a very fast speed (usually at 60 Hz) alternately, so that when the view for the left eye is emitted from the 3D display, only the glasses of the left eye are on, while in the next phase the right eye glasses are turned on and the 3D display emits the view for the right eye. NVIDIA, a pioneer in visual computing technologies, is among the biggest supporters of shutter glasses-based active stereoscopy technology and manufacture GPUs that support active stereoscopy for desktop LCD monitors. Other giant LCD display producers, including Panasonic and Sony, produce active stereoscopic 3DTV sets.

Despite its market support from giant display manufacturers and vendors, active shutter glasses-based 3D stereoscopic display systems also have a number of technical shortcomings. Since the time between the shutter on and off states is quite short, an unpleasant cross-talk effect may become apparent, which is caused by exposing an eye to some extent to the view of the other eye. In order to mitigate this problem, a black image flashes on the screen in every transition from the left-eye view to the right-eye view, and vice versa. But this cure comes at the expense of introducing unnatural flickering into the seen stereoscopic 3D video. Another side effect of the resultant flickering is that it can lead to eye fatigue in long-term video watching and sometimes even dizziness and headaches. This kind of a side effect is less apparent in passive stereoscopic 3D solutions instead. Another disadvantage of watching active stereoscopic displays is that one has to wear heavy and usually bulky active shutter glasses that are rather uncomfortable and tiring.

5.4.4 Auto-Stereoscopic Displays

Auto-stereoscopic 3D displays are able to provide 3D depth perception without any need for specialized eye-glasses as in the case of passive/active stereoscopic 3D display technologies. Thus, this kind of display is also usually called "glasses-free 3DTV". Because of its major advantage of not forcing users to wear those glasses, auto-stereoscopic 3D display technology has also attracted the attention of other consumer electronics developers, such as digital camcorders and mobile video game consoles. Auto-stereoscopic displays basically create the illusion in the brain, so that a planar display can display a 3D image (i.e. an image moving perpendicularly to the screen) by providing a stereo parallax view to the viewer. In other words, both the left eye and the right eye see a different view as in the cases of glasses-based display technologies, which have been calculated to appear from the positions of the two eyes.

The majority of flat-screen auto-stereoscopic 3D displays employ lenticular lens arrays, or parallax barriers, which refract the display image coming from the graphics processor into several viewing directions. While doing so, the full display-resolution video is divided into pieces, equal to the number of resultant viewpoint directions. Thus, each eye sees a reduced-resolution view. When the viewer's head moves in a certain direction, a slightly different view is seen by each eye, creating the overall effect of movement parallax. In early auto-stereoscopic 3D displays, only a single view stereoscopic video was created, which was missing the movement parallax effect.

Auto-stereoscopic 3D displays usually have multiple viewing zones, which results in the fact that multiple viewers can observe the display at the same time, seeing the scene from different viewing angles. However, it is essential that viewers adjust their head positions finely in order to align with the exact ray pattern. This is usually called as the display's sweet spot, where viewers experience a realistic stereoscopic video effect. On the other hand, auto-stereoscopic 3D displays may also exhibit dead zones, where only non-stereoscopic, or no video can be seen.

In parallax barrier-based auto-stereoscopic 3D displays, a mask constructed from a single layer of liquid crystal is laid on the LCD display that directs the emitted light-rays from alternate pixel columns to each eye. The illustration presented on the left side in Figure 5.16 depicts the parallax barrier system. Since it is possible to turn on and off the liquid crystal barrier electronically, it is advantageous to the display user to quickly switch between a 2D and 3D viewing modes.

It should be noted that a particular pixel pattern needs to be applied to the rendered display video, so that when the refracted light-rays arrive at the eyes, the correct imagery is obtained. In lenticular lens set-based auto-stereoscopic 3D displays, an array of cylindrical lenses that is laid on the LCD display directs light-rays from alternate pixel columns to a defined viewing zone. The illustration presented on the right side in Figure 5.16 depicts the cylindrical lenticular lens system.

Multi-view auto-stereoscopic 3D displays bring a new challenge of rendering the multiple views throughout the whole viewing range of the display. Current 3D/multi-view video coding standards tend to restrict the number of original camera views (accompanied by associated depth maps) to be encoded and delivered to display systems. The primary reason for this trend is to afford the transmission/broadcast of suitable 3D media to users. Nevertheless, the multi-view rendering techniques to be employed on the graphical processing units of such auto-stereoscopic multi-view display sets need to

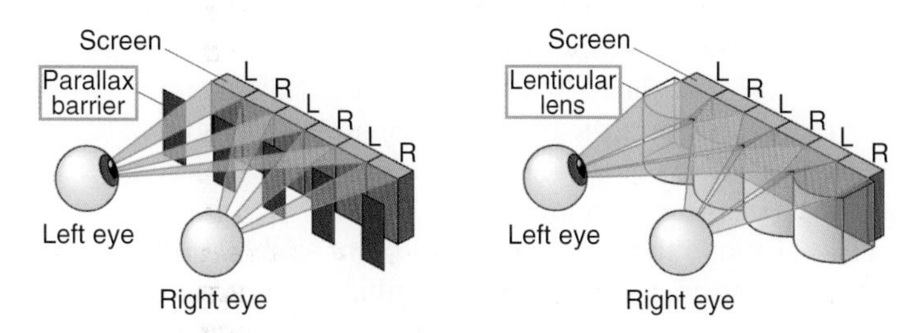

Figure 5.16 Methods of operation for different auto-stereoscopic 3D display technologies (left: parallax barrier, right: lenticular lens)

be improved much further to allow for real-time multi-view playback from two or three view input streams.

The current multi-view auto-stereoscopic displays on the market, which can create from 9 up to 27 viewpoints, are mostly operable with ready-to-display, compatible raw content in real time (i.e. pre-generated viewpoints), but not with two or three camera-plus-depth pairs. Therefore, the 3D market is currently, and will continue for some time to be, dominated by stereoscopic 3D display technologies based on passive-polarized and active-shutter glasses technologies, while the multi-view-plus-depth rendering efficiency and performance for auto-stereoscopic multi-view displays will continue to improve over time.

5.4.5 Light-Field Display

Active and passive stereoscopic displays as depicted in Sections 5.4.3 and 5.4.2 are the most commonly available and sold 3D display systems that are on the consumer market. However, they can only provide the depth feeling from a single and fixed perspective. In other words, regardless of the observation point with respect to the display, every subject is exposed to the same pair of stereoscopic images. As described in Section 5.4.4, the principle of auto-stereoscopic displays is to direct different stereoscopic image pairs to different angles from the screen surface to create motion-parallax effect, but the most widespread types, such as the systems based on lenticular sheet or parallax barrier, have limited spatial resolution.

Viewers are offered a sufficiently smooth and continuous view flow only in a narrow Field of View (FoV). FoV is a crucial feature of any 3D display technology. It refers to the emission range of a display screen. Considering the cone, which is formed by the beams emitted from a particular spot on the screen, the angle (openness) of it refers to the resultant FoV. Similarly, the number of independent light-beams within the FoV that corresponds to the angular resolution, determines the Field of Depth (FoD). Light-field display technology is capable of creating 3D images with quasi-continuous motion parallax in a much wider FoV as opposed to auto-stereoscopic displays.

To display 3D image sequences properly in a realistic sense, a display should produce and emit sufficiently many points in a given time duration [13]. Since the FoV associated with 3D light-field displays is usually larger than that of plano-stereoscopic displays, the number of source (input) views that drive the display to be processed is also larger, which inherently adds to the overall processing workload of the display.

Volumetric displays that constitute a type of such 3D displays create 3D images through emitting or scattering the illumination from a defined point in a three-dimensional space, rather than from a point on a planar surface. A group of modern volumetric displays consist of non-static components, such as the visualization area, which could be a fast rotating mirror, or any special surface that is reflective, transmissive, or a combination of both. The 3D imagery is perceived as a result of the human persistence of vision, such that a remarkably large number of points in the visualization space (regarded as voxels) are illuminated over a unit time frame on the rotating mechanism, where these are integrated into one reconstructed 3D scene in the human brain. Another group of volumetric displays create the 3D images without moving components in the image volume. The volume of space, in which the 3D imagery is produced, is made up of active elements. These elements (voxels) become either completely transparent or opaque/semi-luminous, depending on the rendered 3D scene. Non-tangible visualization medium implementations in this group have gained attention. Gaseous visualization media (e.g. fog), subjected to multiple light projectors can render 3D images, resulting in a volumetric display with static volume that also allows direct interaction with the visualization medium.

Holographic displays are another kind of light-field 3D displays allowing large and continuous FoV 3D imagery. Holography is a technique that enables a light field, which is a product of a light source scattered from objects, to be recorded and later reconstructed in the absence of the original light-field source. Holographic display systems aim at generating holographic patterns to reconstruct the appropriate wavefront. In other words, complex 3D objects can be reproduced from a flat, two-dimensional screen with a complex transparency representing amplitude and phase values [13].

Holographic displays exhibit the cues such as accommodation and blur in the retinal image fully, which are needed to offer viewers an experience free of visual fatigue and eyestrain. A variety of modern digital holographic displays feature a Spatial Light Modulator (SLM). SLMs are objects, which impose some form of spatially varying modulation on a beam of light. SLMs that are based on liquid crystal material should have sufficiently large number of pixels, smaller pixel period, better optical efficiency, and faster operation to realize dynamic and real-time electro-holography [14]. Ideally, an SLM pixel should modulate the amplitude as well as the phase of the refracted light beam in order to properly represent diffraction fields, although the current SLM technology is not able to address both simultaneously. A number of SLM-based holographic 3D imagery solutions have been proposed and presented in the literature [14].

A particular electro-holographic display system generates light rays in its associated optical modules directed at specific points on a screen with various angles, where the exiting angle of light beams is also determined. The screen is in charge of performing the required optical transformation to convert the exposed light beams into a continuous view, while it has a direction selective property with angularly dependent diffusion characteristics [15]. Hence, the optical modules are not necessarily associated with the reconstruction of only particular regions of a real-object, but each projected 3D image can be considered the sum of light beams projected from multiple optical modules behind the screen.

Various light-field display technologies are still in development, and have yet to reach the level of widespread commercialization. A variety of systems developed and prototyped are in use mostly in academic and research institutes and labs, or corporations dealing with specialized 3D imaging applications. The reason why light-field display technologies will not become widely available in the near future is twofold. The first reason is the complexity of the devices necessitating the use of specialized optical units and light sources that inherently is beyond the affordability levels in a potential consumer market. The other, more significant reason is the limited availability of suitable source content that is sufficient to drive the light-field displays resulting in 3D imagery with wide FoV and high spatial resolution. The size of the source multi-view content is large, which also makes it unsuitable for transmission/broadcasting using the existing infrastructure.

References

[1] Shum, H.Y. and Kang, S.B. (2000) 'A review of image-based rendering techniques', in *Proceedings of Visual Communications and Image Processing (VCIP)*, pp. 2–13.

[2] Lang, M., Hornung, A., Wang, O., Poulakos, S., Smolic, A. and Gross, M. (2010) 'Nonlinear disparity mapping for stereoscopic 3D', *Proceedings of ACM Transactions Graph. (Proc. SIGGRAPH)*, **29** (4), 1–10.

[3] Zilly, F. Muller, M., Eisert, P. and Kauff, P. (2010) 'The stereoscopic analyzer–an image-based assistance tool for stereo shooting and 3D production', in *IEEE International Conference on Image Processing (ICIP 2010)*, pp. 4029–4032.

[4] EU ICT ROMEO Integrated Project (Remote Collaborative Real-Time Multimedia Experience over Future Internet), http://www.ict-romeo.eu.

[5] Doyen, D. (2013) 'Interim report on 3D audio video rendering algorithms', Tech. Rep., D3.3, Remote Collaborative Real-Time Multimedia Experience over the Future Internet ROMEO Integrated Project, January.

[6] Boev, A., Bregovic, R. and Gotchev, A. (2010) 'Measuring and modeling per-element angular visibility in multiview displays', Special Issue on 3D displays, *Journal of Society for Information Display*, **26**, 686–697.

[7] Nur, G., Dogan, S., Arachchi, H.K. and Kondoz, A.M. (2010) 'Impact of depth map spatial resolution on 3D video quality and depth perception', in *IEEE 3DTV Conference: The True Vision: Capture, Transmission and Display of 3D Video*, Tampere, Finland, June.

[8] Hewage, C., Worrall, S., Dogan, S. and Kondoz, A. (2009) 'Quality evaluation of colour plus depth map based stereoscopic video', *IEEE Journal of Selected Topics in Signal Processing: Visual Media Quality Assessment*, **3**, 304–318.

[9] Wang, Y.-K., Hannuksela, M. and Pateux, S. (2007) 'System and transport interface of SVC', *IEEE Transactions on Circuits and Systems for Video Technology*, **17**, September.

[10] Nur, G., Dogan, S., Arachchi, H.K. and Kondoz, A.M. (2010) 'Assessing the effects of ambient illumination change in usage environment on 3D video perception for user centric media access and consumption', in 2nd International

ICST Conference on User Centric Media (UCMedia), Palma de Mallorca, Spain, September.

[11] Nur, G., Arachchi, H.K., Dogan, S. and Kondoz, A.M. (2011) 'Advanced adaptation techniques for improved video perception', *IEEE Transactions on Circuit and Systems for Video Technology (TCSVT)*, June.

[12] Tizon, N. (2012) 'Report on streaming/broadcast techniques for 3D multi-view video and spatial audio', Tech. Rep., D4.2, Remote Collaborative Real-Time Multimedia Experience over the Future Internet ROMEO Integrated Project, December.

[13] Balogh, T. (2006) 'The HoloVizio system', *Proceedings of SPIE 6055, Stereoscopic Displays and Virtual Reality Systems XIII*, January 2006.

[14] Reichelt, S., Haussler, R., Leister, N., Futterer, G., Stolle, H. and Schwerdtner, A. (2010) 'Holographic 3-D displays: Electro-holography within the grasp of commercialization', in *Advances in Lasers and Electro Optics*, ed. N. Costa and A. Cartaxo, published under CC BY-NC-SA 3.0 license.

[15] Onural, L., Yaras, F. and Hoonjong, K. (2011) 'Digital holographic three-dimensional video displays', *Proceedings of the IEEE*, **99**, 576–589.

6

Quality Assessment

It is currently difficult to evaluate the perceptual quality of 3D applications, services and the suitability of 3D video coding algorithms for compression and transmission, without resorting to full subjective tests. These subjective evaluation test campaigns take a long time and a great deal of effort to measure the quality using human observers. This has some negative effects on the development and the advance of 3D video broadcasting technologies and new 3D services. The effect of different artefacts introduced by image capture, processing, delivery and display methods on the perceived quality of video are diverse in nature. Video quality models define the relationship between the physical parameter (e.g. coding and delivery method) of the system and the perceived video quality. The objective quality metrics/models that incorporate perceptual attributes for conventional 2D video are well exploited in the literature [1, 2].

This chapter discusses metrics for 2D video quality, and then examines the issues that affect 3D video quality. Methods of measuring 3D video quality using panels of viewers are described in Section 6.3. Finally, details are given of a method for evaluating 3D video quality that correlates well with the HVS. This chapter is based on the work of Dr Lasith Yasakethu, who developed the described quality model during his PhD studies at the University of Surrey.

6.1 2D Video Quality Metrics

Decades of effort on the development of techniques to accurately model how the HVS perceives impairments in video using objective properties have resulted in a large number of 2D image/video quality measurement

3DTV: Processing and Transmission of 3D Video Signals, First Edition.
Anil Fernando, Stewart T. Worrall and Erhan Ekmekcioğlu.

techniques. One of the most extensively used techniques is the Peak-Signal-to-Noise-Ratio (PSNR). The Structural Similarity Index (SSIM) and the Video Quality Metric (VQM) are the other two main metrics widely used in 2D video quality measurements.

6.1.1 Peak-Signal-to-Noise-Ratio (PSNR)

PSNR is derived by finding the Mean Squared Error (MSE) in relation to the maximum possible value of luminance. For an n-bit value, it is defined as follows:

$$PSNR = 20 \cdot \log_{10}\left(\frac{2^n - 1}{\sqrt{MSE}}\right) \tag{6.1}$$

Where $x(i,j)$ and $y(i,j)$ are the original and processed signals at pixel (i,j) and M, N are the picture dimensions. The resultant is a single number expressed in decibels (dB).

Due to the negligible computation complexity associated with the PSNR calculation, it was considered the de-facto standard for measuring image quality for a long time. However, it does not sufficiently correlate with perceptual image quality [3]. For this reason, a number of alternative image quality metrics have been proposed [3].

6.1.2 Structural Similarity Index (SSIM)

Zhou Wang et al.'s method [4] differs from error-based methods in that it uses the structural distortion measurement. The justification for this approach is that the HVS is highly specialized in visualizing structural information from the viewing field rather than visualizing pixel errors. If $x = \{x_i | i = 1, 2, \ldots, N\}$ is the original signal and $y = \{y_i | i = 1, 2, \ldots, N\}$ is the distorted signal, where i is the pixel index, the structural similarity index can be calculated as follows:

$$SSIM = \frac{(2\bar{x}\bar{y} + C_1)(2\sigma_x y + C_2)}{(\bar{x}^2 + \bar{y}^2 + C_1)(\sigma_x^2 + \sigma_y^2 + C_2)} \tag{6.2}$$

In this equation $\bar{x}$ and $\bar{y}$ denote the mean of signal x and y respectively. Variances of x and y are denoted by σ_x and $\sigma_x.\sigma_y$ estimates the covariance of x and y. C_1 and C_2 are constants. The SSIM metric provides an output that falls between 0 and 1. The best quality value of 1 occurs if $x_i = y_i$ for all values of i. For more information about SSIM calibrations and quality evaluation, readers should refer to [4].

6.1.3 Video Quality Metric (VQM)

The Video Quality Metric (VQM) was developed by the Institute of Telecommunication Sciences (ITS) and the American National Standard Institute (ANSI) to provide an objective measurement for perceived video quality.

VQM is a standardized method that closely predicts the subjective quality ratings obtained from a panel of viewers [5]. VQM measures the perceptual effects of video impairments including blurring, jerky/unnatural motion, global noise, block distortion and colour distortion, and combines them into a single metric. Due to its excellent performance in the Video Quality Experts Group (VQEG) Phase II validation tests [6], the ITS VQM methods were adopted by ANSI as a US national standard and as ITU Recommendations in 2004. For more information about the VQM calibrations and quality evaluation techniques, readers should refer to [5].

6.2 3D Video Quality

Unlike conventional 2D video quality, perceived 3D video quality is multidimensional in nature. Even though there are 2D objective quality models that are highly correlated with the HVS, very little work has been carried out on modelling perceptual quality attributes of 3D video. 3D video quality can be considered a combination of a number of attributes such as image quality, depth quality and visual comfort (e.g. eyestrain and visual fatigue). The visual experience of 3D video content can be disturbed by impairments introduced during acquisition, compression, transmission, and reproduction. Traditionally, consumers' visual experience of video content is attributed to the quality of the 2D content. It is defined in terms of the amount and type of distortion introduced by content processing, such as compression artefacts and transmission-related losses.

2D objective quality measures of individual views of a 3D video may not represent the actual 3D video quality as perceived by human viewers. Although some researchers currently use the metrics mentioned above, such as PSNR, to assess 3D video quality, the limitations of PSNR for 2D video have been demonstrated in tests by the Video Quality Experts Group (VQEG) [6]. These limitations are likely to be similar for the assessment of perceptual quality in 3D video content as well. Therefore, explorative studies based on appreciation-oriented methods are required for a true understanding of 3D perception. When 3D video transmission takes place in an error-prone scenario (e.g. terrestrial television broadcasts, or streaming over the internet), objective quality metrics are essential to ensure that the necessary actions along the delivery chain are taken to improve the quality of the transmitted and received data. Examples of such actions include pre-filtering, optimal bit assignment algorithms and error concealment methods. Thus, the availability of objective quality models to predict the attributes of 3D video will speed up the development of 3D broadcast technology, 3D video applications, services and consumer products.

When 3D video is considered, however, image quality alone does not sufficiently represent the quality of the content. It is necessary to assess the depth reproduction accuracy as well. Combining the above two factors, the

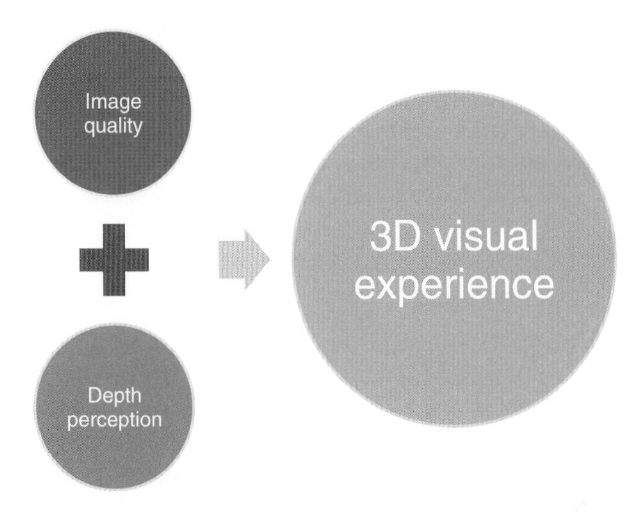

Figure 6.1 3D visual experience

notion of visual experience can be defined as shown in Figure 6.1. The factors affecting perception of image quality and depth are examined in Sections 6.2.1 and 6.2.2 respectively.

6.2.1 Image Quality

The human visual system can identify various types of impairments in video contents. Some of them are spatial while others are temporal impairments. Spatial impairments include blockiness, ringing, contouring, posterizing, and structural discontinuities. Jerkiness, freezing, and mosquito noise are a few examples of temporal impairments. The most accurate way of quantifying the impact of these impairments on the visual experience is by subjective assessments. These assessments involve human subjects, who grade the impaired video [7]. However, subjective assessments are too inconvenient, expensive, and time-consuming for most applications.

6.2.2 Visual Perception of Depth

The human visual system uses both physiological and psychological cues to comprehend the depth in sensed images. Some physiological cues require both eyes to be open (binocular), others are available also when looking at images with only one open eye (monocular). It should be noted that all the psychological cues are monocular. In the real world the human visual system automatically uses all the available depth cues to perceive the depth. The availability of binocular depth cues in a video depends entirely on the capabilities of the particular video reproduction system in use. Nevertheless, monocular depth cues can be reproduced on any existing 2D displays. The physiological depth cues are accommodation, convergence, binocular parallax, and monocular movement parallax. Convergence and binocular parallax are the binocular depth cues while all the others are monocular cues. The psychological depth cues are retinal image size, linear perspective, texture gradient, overlapping, aerial perspective, and shades and shadows.

6.2.2.1 Monocular/Psychological Depth Cues

Image Size

Previous experience allows us to know many objects' relative size. This experience informs our interpretation of distance. The familiar size cue tells us that the visual angle of objects becomes smaller with distance, allowing us to calculate the probable depth or distance of objects. Known size, together with perspective and texture effects, are strong depth cues. When we know that one object is similar in size to another object and both objects are within our plane of vision, the relative size cue allows us to understand that the object with the larger visual angle on the retina is closer than the other object. Our brain then compares the sensed size of the object to this real size, and acquires information about the distance of the object.

Linear Perspective

When looking down a straight level road, one can see that the parallel sides of the road meet in the horizon as illustrated in Figure 6.2. This effect is often visible in photos and it is an important depth cue. It is called linear perspective. Linear perspective frequently occurs with the observation of parallel lines. Such lines appear to recede and converge at the horizon. The horizon appears to rise. This cue is related to the relative size and texture gradient cues, and often all three work together.

Figure 6.2 Linear perspective

Texture Gradient

The closer we are to an object, the more detail we can see of its surface texture. This effect is shown in Figure 6.3. Therefore, objects with smooth textures are usually interpreted as being farther away. This is especially true if the surface texture spans the distance from near to far, with respect to the camera.

Overlapping

When objects block each other out of our sight, we know that an object that obscures another one is closer to us. The object whose outline pattern looks more continuous is felt to lie closer. This effect is illustrated in Figure 6.4.

Aerial Perspective

Aerial perspective is also referred to as relative height. With this cue, the HVS assumes that objects closer to the line of horizon are further away. This phenomenon relates to the way light scatters in the air, causing objects on

Figure 6.3 Texture gradient

Figure 6.4 Overlapping

Figure 6.5 Aerial perspective

the horizon to appear in faded colours or washed in lower light luminance and contrast, while objects that are near appear to have vibrant or intense colours with strong contrast. For example, Figure 6.5 shows the mountains in the horizon always look slightly bluish or hazy. The reasons for this are small water and dust particles in the air between the eye and the mountains. The further away the mountains, the hazier they look.

Shades and Shadows

The shadow monocular depth cue has several rules:

- If an object is solid, it casts a shadow.
- If there is only one light source, then all shadows fall in the same direction and the shadow is opposite from the source of light.
- Objects with shadows falling on them are further away than objects casting the shadow.
- If the object is lower than the 'ground plane' (such as a well), the shadow appears on the same side as the source of light.

When we know the location of a light source and see objects casting shadows on other objects, we learn that the object shadowing the other is closer to the light source. As most illumination comes from above, we tend to resolve ambiguities using this information. The three-dimensional looking computer user interfaces are a nice example of this. Also, bright objects seem to be closer to the observer than dark ones. It is clearly visible in the Figure 6.6.

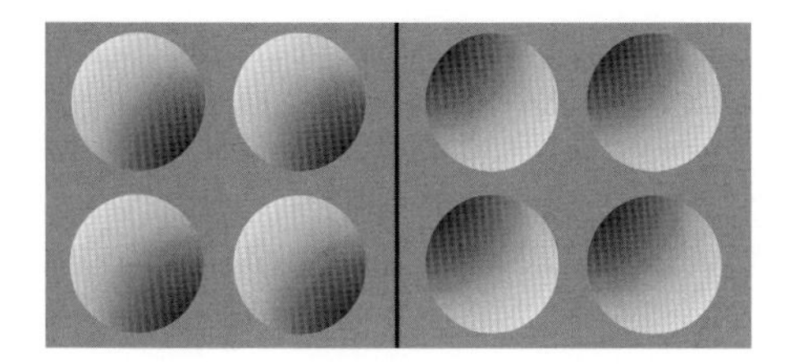

Figure 6.6 Shades and shadows

Monocular Motion Parallax

If we close one of our eyes, we can perceive depth by moving our head. This happens because the HVS can extract depth information in two similar images presented successively to the eyes, in the same way that it can combine two images from different eyes. Motion parallax occurs when an object travels across the retina of a moving person. The moving person focuses on one object while noting the relative movement of other objects. Parallax causes objects in the distance to appear to move more slowly than objects that are closer. Kinetic depth perception allows us to gauge the velocity of moving objects. When an object moves away, it appears to grow smaller. When an object approaches, it appears to become larger. We constantly judge changes in our positions to others using kinetic depth perception cues.

6.2.2.2 Binocular/Physiological Depth Cues

Accommodation

Accommodation is the tension of the muscle that changes the focal length of the lens of eye. Thus it brings into focus objects at different distances. The muscular activity necessary for this accommodation acts as a signal for the brain to generate perception of depth and distance. This depth cue is quite weak, and it is effective only at short viewing distances (less than 2 metres) and with other cues. The peripheral vision cue occurs due to the curvature of the eyeball. This curvature causes the visual field to distort or appear to bend at its extreme edges. This visual distortion is accommodated for when we interpret an image or scene. Often we ignore that the lines of objects, which our prior knowledge understands are straight, appear to be curved. The effect of this curvature can be seen in some photographs where no accommodation has taken place, which is partially why photographs often do not capture the image we think we have seen.

The accommodation cue occurs when the dioptic power of the lens increases and allows close objects to be focused clearly on the retina. How this cue informs the understanding of distance is not yet clearly understood by scientists.

Convergence

Binocular convergence is based on the fact that in order to project images on the retinas, the two eyes must rotate inward toward each other. The closer the perceived object is, the more they must rotate, so the brain uses the information it receives about the degree of rotation as a cue to interpret the distance of the perceived objects. This depth cue is effective only for short distances (less than 10 metres).

Binocular Parallax

As our eyes see the world from slightly different locations, the images sensed by the eyes are slightly different, as shown in Figure 6.7. This difference in the sensed images is called binocular parallax. The HVS is very sensitive to these differences, and binocular parallax is the most important depth cue for medium viewing distances. A sense of depth can be achieved using binocular parallax even if all other depth cues are removed.

Micro Head Movements

This depth cue is related to movement parallax, but is much less apparent than the previously described motion parallax effects. Neuroscientists' research in the field of human reception of 3D found that binocular parallax in itself is not enough for the brain to understand the 3D space [8]. A brain area, the anterior intraparietal cortex (AIP), integrates binocular and motion parallax [9]. As soon as the brain thinks that it sees a 3D image, it starts working as if in a normal 3D world, employing micro head movements to repeatedly and unconsciously check the 3D model built into our brain. When the displayed 3D image is checked and the real 3D world mismatches the 3D image, the trick is revealed. Presumably the AIP cortex never got used to experiencing such 3D cue mismatch during its evolution, and this mismatch produces unwanted effects.

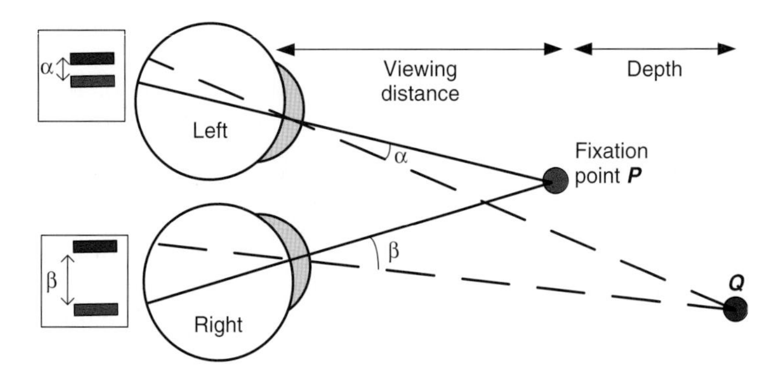

Figure 6.7 Binocular parallax

6.3 3D Video Quality Evaluation Methods

Due to the multi-dimensional attributes of 3D video such as image quality, depth quality and visual comfort, it is currently difficult to evaluate the perceptual quality of 3D applications, services and the suitability of 3D video coding algorithms for compression and transmission, without resorting to full subjective tests. Therefore subjective test campaigns are widely carried out to understand the true quality of 3D representation [3]. Subjective quality evaluation methodologies for stereoscopic television pictures are described in ITU-R BT.1438 [10]. Subjective evaluation procedures in this recommendation are based on the ITU-R BT.500.11 quality assessment recommendation for television pictures [11]. The main assessment methodologies considered in [10] and [11] are described in Table 6.1.

DSCQS and SSCQS methods are used for subjective assessment of 3D video, as these methods are recommended by standardization bodies for 3D video quality measurements and are widely used in 3D video research [12, 13]. The presentation structure of test materials and the quality-rating scales for both DSCQS and SSCQS methods are illustrated in Figure 6.8 and Figure 6.9 respectively. In the DSCQS method, reference and processed

Table 6.1 Assessment methodologies for subjective testing

Assessment methodology	Description
Double-Stimulus-Continuous-Quality-Scale (DSCQS) method	Measures the quality of the processed (or distorted) video sequence compared to its undistorted reference (or original video sequence). The assessor provides a quality index for both processed and undistorted video sequences.
Single-Stimulus-Continuous-Quality-Scale (SSCQS) method	A single video is presented without its reference and the assessor provides a quality index individually for each video sequence in the stimulus set.
Threshold estimation method	Series of video sequences are presented sequentially in time. The assessor is asked to assess the point at which impairment becomes visible.
Stimulus comparison method	Series of video sequences are presented sequentially in time. The assessor is asked to assign a relation between two consecutive video sequences.

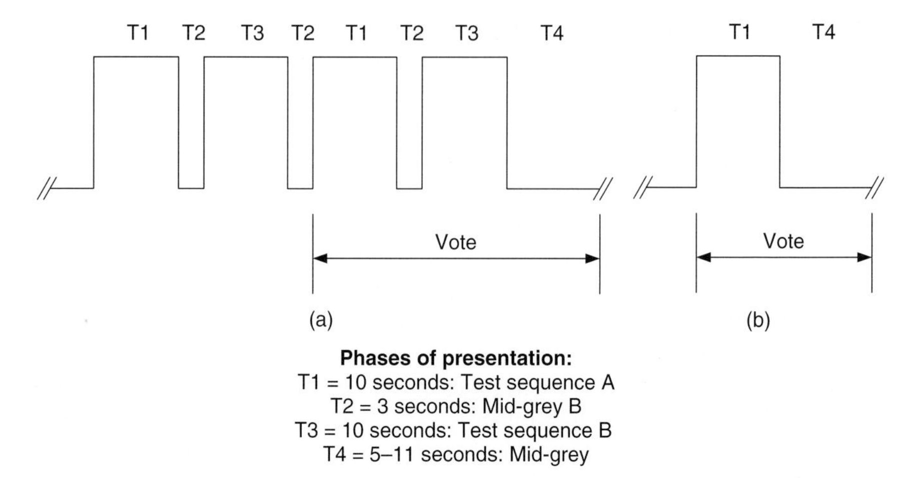

Figure 6.8 Presentation structure of material. (a) DSCQS method (b) SSCQS method

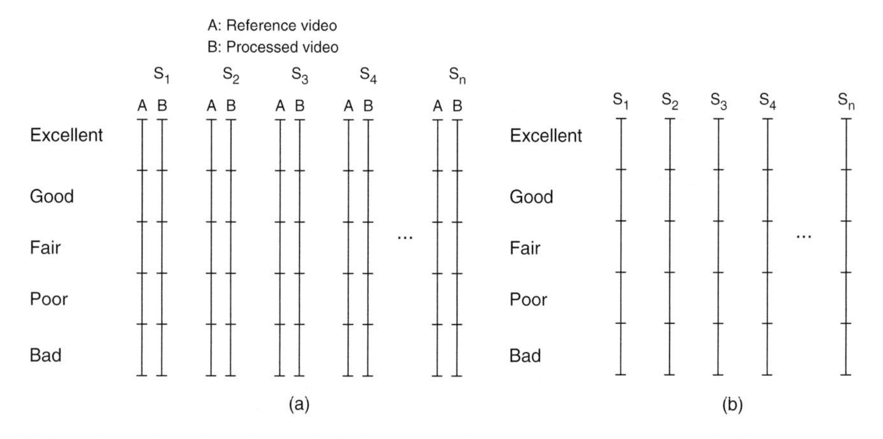

Figure 6.9 Quality rating scale for subjective assessment, where S_n represents video sequence *n*. (a) DSCQS method (b) SSCQS method

videos are presented twice and the quality rating is calculated as the difference in the mean opinion scores (MOS) between the reference and the processed video sequence.

Subjective evaluation test campaigns, using human observers, take a significant amount of time and effort. Moreover, they cannot be used in on-the-fly quality optimization algorithms (e.g. rate-distortion optimization). This has negative effects on the development and advance of 3D video broadcasting technologies and new 3D services. Therefore, there is a clear requirement for objective quality methodologies to predict the quality of

3D video. Objective 3D quality assessment is still far from being mature. Objective video quality assessment (for both 3D and 2D) is generally classified as follows:

- Full-reference objective video quality assessment (FR);
- No-reference objective video quality assessment (NR);
- Reduced-reference objective video quality assessment (RR).

In FR video quality assessment it is assumed that the quality assessment technique has full access to both processed and original versions of the video sequence. The typical functions of a FR assessment model are described in Figure 6.10(a). In some cases, such as video quality evaluation on the receiver side, FR methods may not be applicable as the reference video sequence is not available. On the other hand, NR video quality assessment uses only the information from the processed video to produce an objective quality score. These models do not have access to the reference video. However, they use prior information about the distortion of the video sequence. For instance,

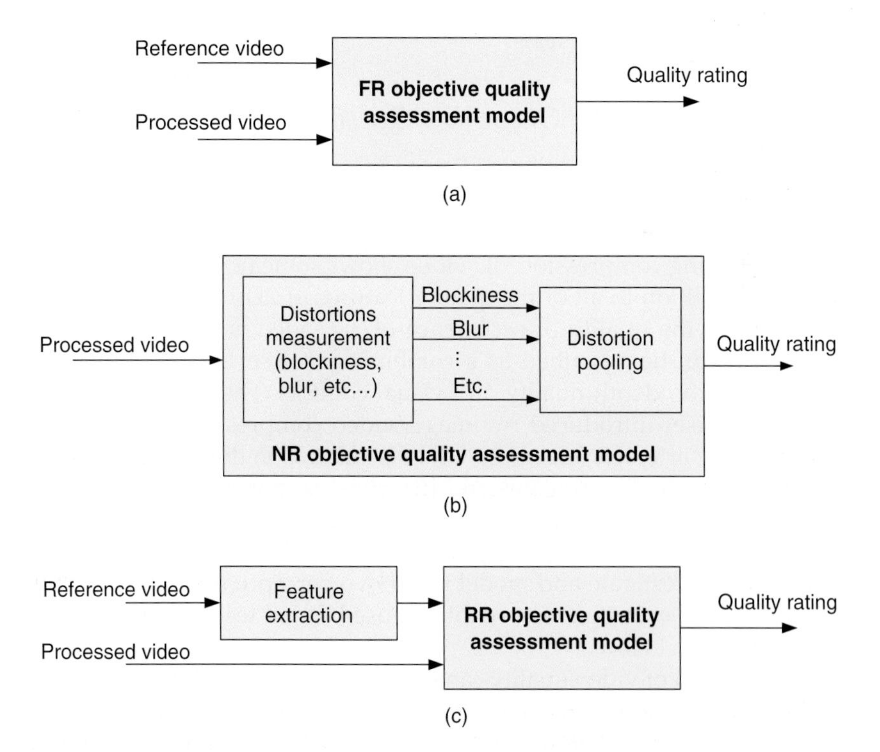

Figure 6.10 Objective quality assessment techniques. (a) FR objective quality assessment model (b) NR objective quality assessment model (c) RR objective quality assessment model

they know the type of encoding scheme that was used, in order to look for codec-specific distortion, such as blockiness, blur, etc. Therefore, typically NR quality models are designed for one or a set of specific distortion types and are unlikely to generalize for assessment of other types of distortion. The NR quality assessment model is described in Figure 6.10(b).

RR video quality assessment provides a solution that lies between FR and NR assessment. In RR quality assessment, certain features that are extracted from the reference video are available for the quality assessment model to evaluate the quality of the processed video. In a practical context, these important features must be coded and transmitted with the compressed video produced by the encoder to the RR quality assessment model. In general, the extracted features are coded using a much lower data rate than the compressed video and are transmitted through an error-free channel or in the same channel but with better protection (e.g. through error control coding) [14]. RR quality assessment is illustrated in Figure 6.10(c).

The goal of objective 3D video quality assessment research is to design effective models to automatically predict the quality attributes of 3D video. This will enable service providers, developers and the standards organizations to rely on meaningful quality evaluation methodologies without resorting to full subjective tests.

6.3.1 Subjective and Objective Quality Measurements

Efficient compression techniques are vital for bandwidth-limited communication channels. Regardless of the coding methods used, rendered 3D video quality is affected by the artefacts introduced to the colour texture and depth sequences during compression. 3D video shows some new classes of quality features in addition to all of the quality features of 2D video. As discussed in Section 6.2, the quality of experience of 3D video is multi-dimensional in nature. It can be described as a combination of several attributes such as image quality, depth quality and visual comfort. The impairments (e.g. blockiness, noise) introduced by image/video compression, produce specific 3D quality artefacts (e.g. cross-talk, discomfort) in the reconstructed 3D video. This chapter investigates and describes an approach for modelling two dominant perceptual quality attributes, i.e. image and depth quality, of 3D video from the user perspective. Several subjective experiments are conducted to investigate and model the above perceptual quality attributes of 3D video. The common test conditions used for all tests described in this chapter are given below.

The evaluation of video quality can be divided into two classes: subjective and objective methods. Intuitively one can say that the best judgement of quality is the human being. This is why subjective methods are said to be the most precise measures of perceptual quality, and to date subjective experiments are the only widely recognized method of accurately determining perceived quality [3]. In subjective experiments, humans are involved to evaluate the quality of a video in a controlled test environment. This can be done by providing a distorted video for evaluation by the subject. Another way is to additionally provide a reference/original video that the subject can use to determine the relative quality of the distorted video. These different methods are specified for television sized pictures by ITU-R [11] and are, respectively, referred to as Single Stimulus Continuous Quality Scale (SSCQS) and Double Stimulus Continuous Quality-Scale (DSCQS) methods. Similarly, for multimedia applications an Absolute Category Rating (ACR) and Degradation Category Rating (DCR) are recommended by ITU-T [15]. Common to all procedures is the pooling of the votes into a Mean Opinion Score (MOS) which provides a measure of subjective quality of the media in the given test set.

Subjective quality assessment has two obvious disadvantages. First, subjective quality assessment is expensive and very tedious as it has to be performed with great care in order to obtain meaningful results. Second, it is impossible to integrate subjective testing into the application development cycle in real time. Hence, automated methods that attempt to predict the quality as it would be perceived by a human observer are necessary. These quality measures are referred as objective quality measures. There are a wide variety of existing methods, from computationally and memory efficient numerical methods, to highly complex models incorporating aspects of the HVS [16].

In the last two decades, a lot of objective metrics have been proposed to assess image/video quality. They can be divided into two main categories, namely, psychophysical methods and statistical methods. Metric design following the former approach is mainly based on the incorporation of various aspects of the HVS that are considered crucial for visual perception. This can include modelling of contrast and orientation sensitivity, spatial and temporal masking effects, frequency selectivity and colour perception. A number of such quality metrics have been proposed by different authors in the literature, such as the Video Quality Metric (VQM) [5], the Perceptual Video Quality Measure (PVQM) [17], and the Moving Picture Quality Metric (MPQM) [2]. Methods following the statistical approach are mainly based on image analysis and feature extraction. These methods also contain certain aspects of HVS parameters as well. The extracted features and artefacts can be of different kinds, such as spatial and temporal information. Ultimately, irrespective of the nature of the objective metric, its outcome can be connected to human visual perception by relating them to the MOS obtained in subjective experiments. Therefore, subjective experiments are still required to validate newly proposed objective metrics. We will now discuss some of the key features of subjective experiments.

6.3.1.1 Subjective Test Design

The experiment has a within subjects design, with several colour-plus-depth-based 3D video sequences and different encoding combinations as independent variables and 3D quality attribute(s) as dependent variable(s).

6.3.1.2 Selection and Screening of Observers

Observers are usually either classified as being expert or non-expert viewers. Expert viewers will usually give more consistent results, but their opinions may not be representative of those of the general public. They are often very good at recognizing artefacts within video compared to non-expert viewers. For this reason, when expert viewers are used, a lower number of observer are used (e.g. 5–10 viewers). If non-expert viewers are employed, then it is necessary to carry out a much larger number of tests to ensure the reliability of results, and to be able to exclude outliers.

For the tests in this chapter, 28 non-expert observers volunteered to participate in all experiments. The observers are mostly research students and staff with a multimedia signal processing background. Their ages range from 20 to 40. All participants had a good visual acuity (>0.7, as tested with a Snellen eye chart), good stereo vision (<60 seconds of arc, as tested with the TNO stereo test) and good colour vision (as tested with the Ishihare test). Each assessor is well informed on the test process and the test materials (possible quality defects).

Eye dominance refers to the tendency of the brain to prefer visual input from one eye over the other. The 'Finger-Point' method is often used to determine eye dominance. In this method, observers are pointed naturally at an object with both eyes open and the face square to the object. The eyes are closed alternately. When the dominant eye is closed, the finger appears to jump away from the original location. Eye dominance is particularly important when evaluating techniques, such as asymmetric compression, where the left and right eyes are compressed with different qualities.

6.3.1.3 Displays

The 42″ Philips WOWvX multi-view auto-stereoscopic display (display resolution-1920 × 1080, aspect ratio-16:9 and the peak luminance of the display is 200 cd/m^2) is used in all experiments to display the 3D content. The advantage of this display, besides 3D viewing without glasses, is the support of motion parallax enabling the viewer to look around objects by moving their head. The viewing distance for the observers is set at 3m, which is optimum for the display optics. The 3D display is calibrated using a GretagMacbeth Eye-One Display 2 calibration device. The measured environmental illumination is 190 lux, which is closer to the recommended value in [11] for home environments (i.e. 200 lux). The background luminance of the wall behind the display is 20 lux. These environmental luminance measures remained the same for all test sessions, as the lighting conditions of the test room are kept constant. The original, MPEG-4 SVC coded colour-plus-depth image sequences are initially combined in side-by-side video sequences. These side-by-side image sequences are then input to the 3D display to allow subjects to evaluate the perceptual quality.

6.3.1.4 3D Video Compression

The encoding parameters used are listed in Table 6.2, and the details of the video sequences considered in this chapter are given in Appendix A.

JSVM version 8.9, developed by the JVT (Joint Video Team), is used to perform SVC compatible encoding of both colour texture and depth video sequences. When encoding colour-plus-depth-based 3D videos, the base layer of the codec is used to encode the colour texture video, and the enhancement layer is used to encode the corresponding depth map of the 3D sequence. Performances of the quality assessment methodologies considered in this chapter are verified for eight video sequences through subjective testing. The stimulus set for each test differs depending on the quality variations of colour texture video and depth map, and is described in the following sections. In each of the subjective evaluation tests, observers are asked to rate the 3D video for the particular quality attribute(s) under investigation according to the Double Stimulus Continuous Quality Scale (DSCQS) method, as described in ITU-Recommendation BT.500-11 [11]. Before each test, a handout is given to the observers explaining about the assessment, the grading scale, the sequence and timing (reference picture, grey, test picture and voting period), according to in ITU-Recommendation BT.500-11. An example handout is shown in Appendix B. Example clips are shown to familiarize the viewer with the assessment procedure. Any remaining questions are answered before the start of the test and they are assisted during the whole test procedure.

Table 6.2 Encoding parameters

Encoding parameter	Value
Encoder	JSVM (Joint Scalable Video Model) reference software codec Version 8.9
Number of sequences	8
Sequence length	10s
Sequence format	IPPP
Reference frames	1
Entropy coding	CABAC (Content Adaptive Binary Arithmetic Coding)

6.3.1.5 Processing of Results

After the subjective test scores have been collected, they can be processed using the procedures set out in [11]. This will require discarding of results from individual observers that are significantly out of line with the average results. It is also necessary to calculate confidence intervals to show whether the differences in the quality assessments are statistically significant.

When the subjective results are being used to validate a newly proposed objective metric, there are further procedures to follow. The relationship between the above objective quality assessment models and the subjective quality ratings for user perceived image quality (i.e. MOS) may be approximated by a symmetrical logistic function Equation (6.3), as described in ITU-R BT.500-11 [11]:

$$p = \frac{1}{[1 + e^{(D-D_M)\cdot G}]} \tag{6.3}$$

where, p is the normalized opinion score, D is the distortion parameter, D_M and G are constants and G may be positive or negative. The quantitative measures for each prediction model approximated using the symmetrical logistic function are presented in Table 6.3. The Pearson Correlation Coefficient (CC), Root Mean Squared Error (RMSE) and Sum of Squares due to Error (SSE) are used as performance comparison metrics to evaluate the objective quality models. It should be noted that, CC = 1, RMSE = 0, SSE = 0 and CC = 0, RMSE = 1, SSE = 1 will indicate perfect and worst correlation between the attribute considered and the objective model respectively.

Table 6.3 Performance of image quality assessment models for 3D video sequences

Objective quality model		Image quality		
		CC	RMSE	SSE
PSNR	Measure 1	0.8808	0.0696	0.06226
	Measure 2	0.8811	0.06957	0.06223
	Measure 3	0.8885	0.06881	0.06148
	Measure 4	0.8917	0.06763	0.06041
SSIM	Measure 1	0.8424	0.1798	0.09934
	Measure 2	0.8484	0.1708	0.09026
	Measure 3	0.8692	0.1527	0.07934
	Measure 4	0.936	0.04241	0.0439
VQM	Measure 1	0.9433	0.03681	0.04174
	Measure 2	0.9189	0.06201	0.05568
	Measure 3	0.9321	0.04483	0.04514
	Measure 4	0.96	0.03506	0.03992

6.3.2 Effects of Colour Texture Video and Depth Maps on Perceptual Quality

In order to quantify the effects of the quality of colour texture (monocular) video and the depth map (disparity signal) on image and depth quality, two experiments are conducted. Image quality refers to the spatial and temporal texture quality of the 3D content. However, the image quality alone does not sufficiently represent the overall quality of the 3D content. It is necessary to assess the accuracy of the depth reproduction as well. This notion is identified as the depth quality.

In the first experiment (test 1), the quality of the colour texture video is varied by different quantization settings, (quantization parameter = 10, 20, 30, 35, 40, 45, 50) while the depth map is not compressed. In the second experiment (test 2), the quality of the depth map is varied using the same QP values employed on the colour video in test 1. The colour texture video is not compressed for test 2. The stimulus set contains seven impaired video sequences of each scene. The original, uncompressed representation (both colour texture and depth map videos with original quality) is used as the reference in the stimulus.

The test sequences are randomized and presented sequentially for each test and the observers are asked to rate the 3D video for image and depth quality according to the DSCQS method. During the analysis of results, the differences in subjective ratings for the impaired video sequences and the original video sequences are calculated. Then the difference is scaled into a linear opinion score scale, which ranges from 0 (excellent) to 100 (bad). The MOS for each test sequence is obtained after averaging the opinion scores for all subjects. The results are then illustrated in Figure 6.11 for the 'Breakdancers', 'Orbi', and 'Interview' video sequences, considering their different motion characteristics (see Appendix A). Figures show the resultant MOS for the image quality and depth quality, with the different quality levels of 3D content (colour texture or depth map) considered for tests 1 and 2. In Figure 6.11, the error margins are calculated using Standard Error (SE) of mean, for image and depth quality MOS.

When the quality of the colour texture/depth map video is varied, the user experience can be summarized as follows:

> The perceived image quality of 3D video is significantly affected by the quality of colour texture video. Although 3D video shows novel classes of quality features, it also depends on all of the quality features of 2D video. On the other hand, the quality of the depth map has little impact on image quality perception. However, the depth quality of 3D video is affected by the quality variations of both colour texture video and depth map. This is due to the fact that the human brain utilizes various types of visual information known as depth cues available in a scene to build a unified perception of depth. Some depth cues can be perceived with a single

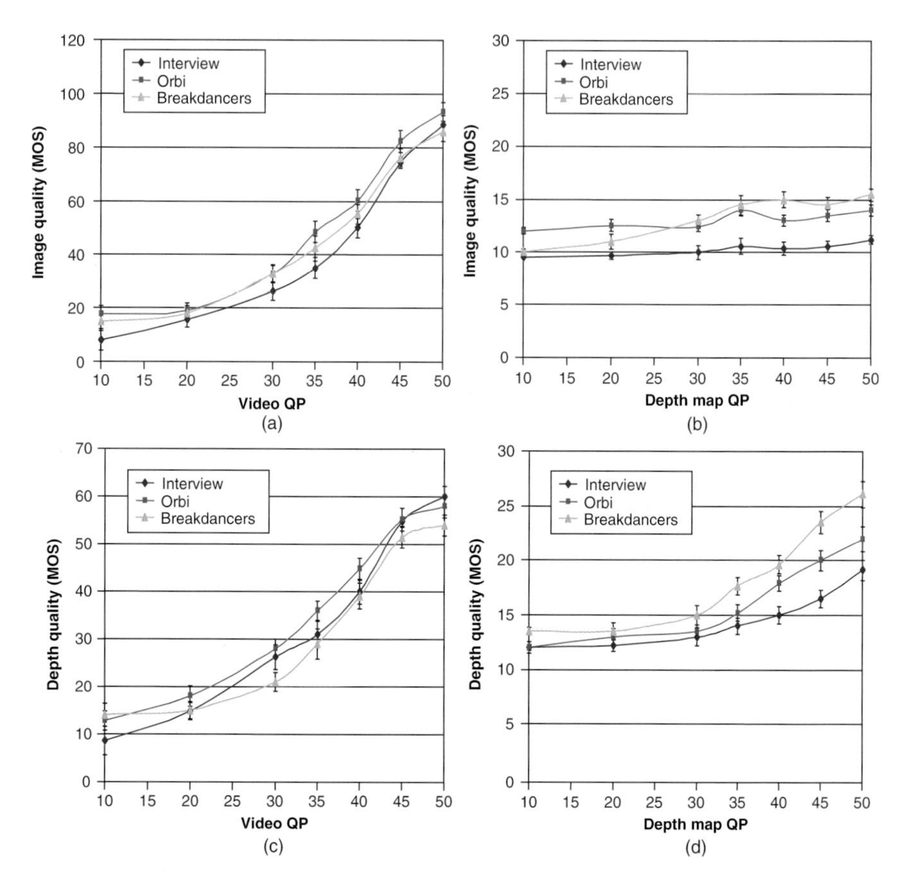

Figure 6.11 Variation of subjective results for the 3D quality attributes. (a) Image quality MOS scores and compression of colour texture video (b) Image quality MOS scores and compression of depth map (c) Depth quality MOS scores and compression of colour texture video (d) Depth quality MOS scores and compression of depth map

eye (monocular cues) such as shadows, perspective and motion parallax. Other cues like stereopsis and convergence (binocular cues) require the cooperative work of both eyes. Thus depth perception of 3D video is affected by both monocular and binocular cues (extracted from the colour texture video and depth map). Also it is noted that the depth quality is more sensitive to quality variation of the colour texture video than the depth map.

6.4 Modelling the Perceptual Attributes of 3D Video

This section attempts to objectively evaluate the quality attributes of colour-plus-depth-based 3D video described in Section 6.2. A third experiment (test 3) is conducted with quality variations to both colour texture and depth map videos. The colour texture video is compressed with QP values of 0, 10, 30, 35, 40, 50 and the depth map with QP values of 0, 10, 35, 40, 50. The compression levels are selected by analysing the variation of MOS attributes (i.e. image and depth quality) from tests 1 and 2. Each impaired colour texture video is coupled with all compressed levels of depth maps, to give all possible combinations of colour and depth quality. All test conditions are similar to those used in tests 1 and 2 (see Section 6.3.2). Subjects are asked to rate the 3D video for image quality, depth quality and overall 3D quality according to the DSCQS method.

6.4.1 Modelling the Image Quality of 3D Video

To objectively evaluate the image quality attribute of 3D video, three widely used 2D objective quality models have been selected. In view of these different quality models, four different combination approaches are considered, incorporating the visual characteristics of observers:

$$Measure\ 1 = \frac{1}{2}[Obj_L + Obj_R] \tag{6.4}$$

$$Measure\ 2 = \frac{1}{N_{tot}}[N_L \cdot Obj_L + N_R \cdot Obj_R] \tag{6.5}$$

$$Measure\ 3 = \frac{1}{N_{tot}}\left[\frac{\sum_{i=1}^{n} Acuity_{L,i}}{Acuity_{Max}} \cdot Obj_L + \frac{\sum_{i=1}^{n} Acuity_{R,i}}{Acuity_{Max}} \cdot Obj_R\right] \tag{6.6}$$

$$Measure\ 4 = Obj_{colour} \tag{6.7}$$

Where, Obj_L and Obj_R are the objective quality ratings of the rendered left and right videos, and Obj_{colour} is the objective quality of the colour texture video. N_{tot} is the total number of observers, N_L and N_R refer to the number of observers whose dominant eye is respectively the left or right. In the tests described here, $N_L = 16$ and $N_R = 12$. $Acuity_{L,i}$ and $Acuity_{R,i}$ represent the visual acuity of the left and right eye of the ith observer respectively and represents the maximum value of visual acuity. PSNR, SSIM, and VQM are considered as objective measures for each of the measurement techniques considered.

In the average approach (Measure 1), objective scores from the rendered left and right videos are averaged. In the dominant eye approach (Measure 2), the dominant eye of the observer is taken into account. This combination tries to weight different objective scores giving more importance to the evaluation of image quality of the video relative to the dominant eye of the observer. The objective metrics evaluated independently on the right and left rendered video of the stereo pair have also been combined by means of the visual acuity approach (Measure 3). Specifically the scores obtained for the left and right image have been weighted by using the visual acuity of the observers. The rationale behind this approach is that the perceived quality could decrease along with the visual acuity of the observer. The maximum visual acuity of the observer is defined to be equal to 1. A Snellen eye chart is used to measure the visual acuity of the test subjects. In Measure 4, the objective quality rating of colour texture video is considered.

Some common objective metrics have been compared to the actual MOS results obtained from subjective experiments. The correlation of the results has been assessed using the procedures described in Section 6.3.1. According to Table 6.3 all quality models are generally acceptable in predicting the image quality attribute of 3D video. However, the results show that Measure 4 of VQM has the best correlation with respect to subjective ratings for predicting image quality. Thus, the VQM quality rating of the colour texture video can be effectively used to predict the image quality of colour-plus-depth-based 3D videos.

6.4.2 Modelling the Depth Quality of 3D Video

From tests 1 and 2, it is noted that depth quality is affected by both colour texture video (contributing to monocular cues) and the depth map (contributing to binocular cues). However, experiments with so-called random-dot stereograms show that binocular and monocular depth cues are independently perceived. In a similar way, human brains will independently perceive monocular and binocular degradation in a 3D video [7, 18]. Furthermore, the added value of strength of depth by the depth map (3D viewing compared to 2D) can be identified as the effect from the superimposition of binocular cues on monocular cues. Thus, the overall depth quality can be modelled as follows:

$$Depth\_quality = f(D_M, D_B) \tag{6.8}$$

$$Depth\_quality = D_M^{\alpha} D_B^{\beta} \tag{6.9}$$

where D_{tot} refers to the objectively evaluated inclusive depth quality and D_M and D_B refer to the contribution to depth quality from monocular and binocular cues respectively. α and β are positive constants that define the relative importance for depth quality. Due to the independent nature of monocular and binocular cues of depth perception, first the contribution from the monocular cues to overall depth quality is evaluated and, second, the added value from the binocular cues is modelled.

6.4.2.1 Contribution from the Colour Texture Video

If no depth map is presented, similar to 2D video, depth quality is judged only from the monocular cues extracted from the colour texture video. A fourth experiment (test 4) is conducted with quality variations in colour texture video with QP values of 0, 10, 30, 35, 40, 45, and 50. Test conditions are similar to the experiments previously described in this chapter. The stimulus set contained impaired colour texture video (i.e. 2D content) and the original, uncompressed version of each scene as the reference.

All pixel values of the corresponding depth map sequences are set to a greyscale value of 128 (according to 3D display characteristics), so that the colour texture video will display on the same plane as the screen of the display. Subjects are asked to rate the level of perceived depth according to the DSCQS method. Here the perception of depth refers to the feeling of depth experienced from the monocular cues of the video. An undistorted colour texture video will preserve the inherited monocular cues of depth in its original video. Thus, the distortion between the original and processed videos will indicate the deterioration of the monocular cues of the colour texture video.

PSNR, SSIM, and VQM are selected as assessment models to predict the perceived depth from monocular cues extracted from the colour texture video. The relationship between the perceived depth (i.e. MOS) and the assessment models are approximated by a symmetrical logistic function as shown in Equation (6.3). The quantitative measures of each prediction model (i.e. PSNR, SSIM and VQM), approximated using the symmetrical logistic function, are presented in Table 6.4. Performance comparison metrics (CC, SSE, RMSE) are evaluated for all test sequences considered. The results imply that VQM has better values for the performance comparison metrics, with respect to subjective rating, in predicting perceived depth from colour texture video. Therefore, VQM can be used to predict perceived depth from monocular cues extracted from the colour texture video.

6.4.2.2 Contribution from the Depth Map

The next stage is to model the added value of the depth quality from the depth map (also referred to as the disparity signal in this section). Existing 2D objective quality models are not suitable to evaluate the depth map for its contribution towards binocular cues (e.g. stereopsis) for the overall

Table 6.4 Performance of depth quality assessment from colour texture (monocular) video sequences

Objective quality model	Depth quality		
	CC	RMSE	SSE
PSNR	0.8226	0.1659	0.09345
SSIM	0.8431	0.1027	0.07352
VQM	0.9129	0.1007	0.0596

depth quality. This is due to the fact that depth maps are not natural images (i.e. depth maps are not directly perceived by the 3D observer). Thus, a new full reference disparity distortion model (DDM) is needed. The measurement technique described here is constructed using three model parameters evaluated between the original depth map and the corresponding depth map processed after degradation.

When experiencing 3D video, the HVS identifies depth through visual recognition of discrete depth planes. The relative distances to different depth planes and the consistency of the content (as compared to the original video) in the identified planes, aid in visualizing the depth of 3D content. Being able to recognize the distance (or separations) between clusters of visual objects assists the 3D observer to identify depth when experiencing 3D video. The relative positions of the objects within the object clusters should be consistent with the original 3D video in order to provide a satisfactory sensation of depth as intended by the original content. To identify the depth of an object, the eyes converge to a region by matching the visual information projected in both retinas.

The first binocular cells that react to a stimulus, presented to either of the eyes, appear at a later stage of the visual pathways known as the primary visual cortex or V1 area of the brain. At this stage, only the structural information extracted separately from each eye is available to the brain for deduction of image disparity [17]. Thus it is believed that HVS is highly optimized in extracting structural information from the viewing field for perception of depth. A simplified diagram of the HVS is shown in Figure 6.12. The Lateral Geniculate Nucleus (LGN) of the brain recognizes the visual streams originated from the eyes before being relayed to the primary visual cortex. Although the function of the primary visual cortex is fairly well understood and accepted by the vision science community, research is still being carried out to understand the higher-level vision processing and cognition of the later stages of human vision [19].

In view of the above observations, important features of HVS for depth perception are considered in designing the proposed objective quality model. The system diagram of the proposed quality assessment system is shown in Figure 6.13. Suppose X and Y are two disparity signals, which have been aligned with each other. If one signal is assumed to have perfect quality (signal X), then the DDM measure can serve as a quantitative measurement of the quality of the second signal Y. The proposed model separates the task of the measurement technique into three components given below:

- M_1 = distortion of the relative distance in depth axis among depth planes
- M_2 = distortion of the consistency in perceived depth of the contents in depth planes
- M_3 = structural comparison.

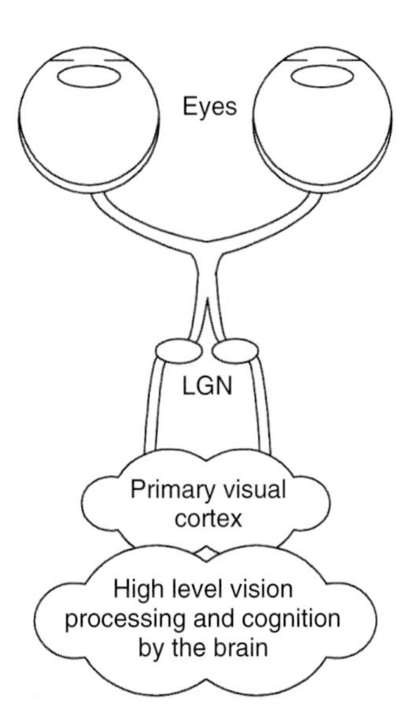

Figure 6.12 Schematic diagram of the HVS

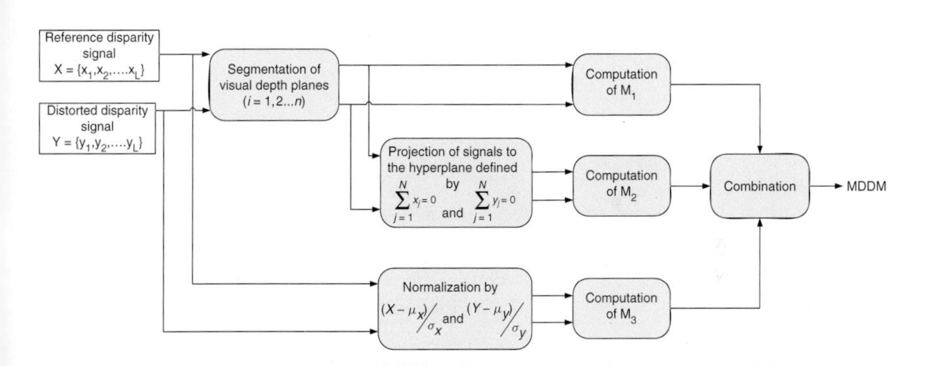

Figure 6.13 Block diagram of the proposed measurement system

In order to identify the visually recognized depth planes, histogram of the undistorted (original) disparity signal is examined. Figure 6.14 to Figure 6.16 show histograms for the sequences 'Breakdancers', 'Orbi', and 'Interview' respectively. Examining the distribution of the per-pixel values of the disparity signal (0-furthest and 255-closest to the camera), visually recognized depth planes are identified as shown in Figure 6.14 to Figure 6.16. After identifying the pixel variation in different depth planes, the depth map (both original and degraded) is segmented accordingly. Resultant depth planes for the 'Breakdancers' sequence are shown in Figure 6.17.

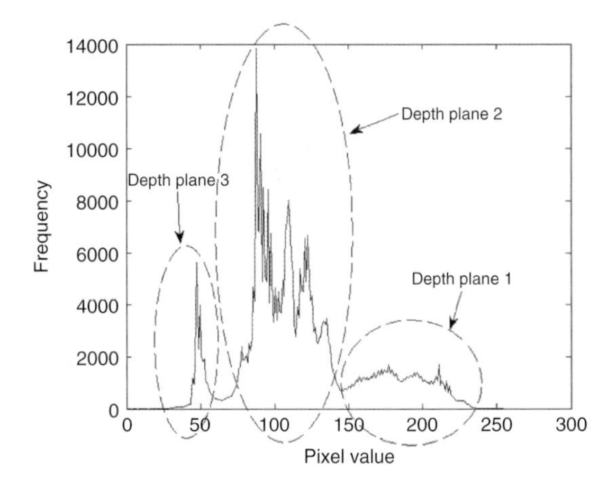

Figure 6.14 Identifying depth planes of Breakdancers sequence

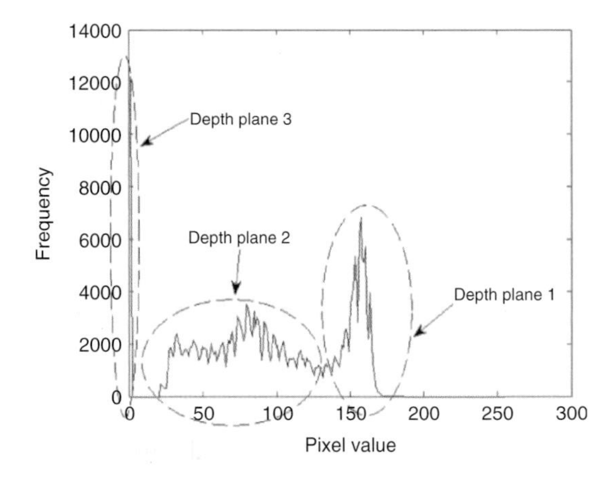

Figure 6.15 Identifying depth planes of the Orbi sequence

Calculating M_1: Distortion of the Relative Distance in Depth Axis Among Depth Planes

The relationship between the actual depth value z and its corresponding unsigned 8-bit pixel value m of the depth map is as follows [20]:

$$z = \frac{m}{255} \cdot (k_{near} W + k_{far} W) - k_{far} W \tag{6.10}$$

Positive and negative values of z mean that the corresponding pixel should be in front and behind the display respectively. k_{near} and k_{far} specify the range

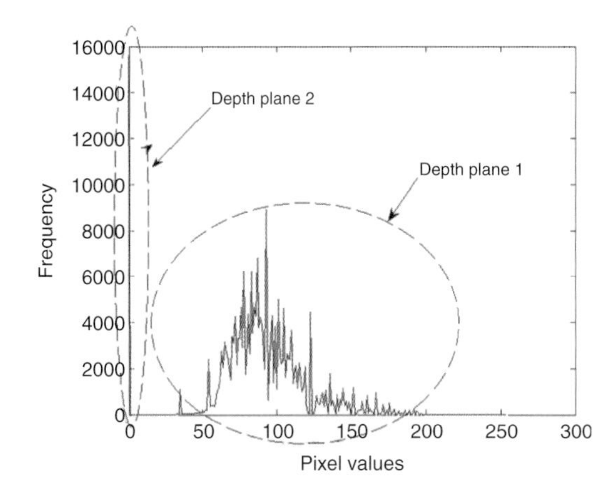

Figure 6.16 Identifying depth planes of the Interview sequence

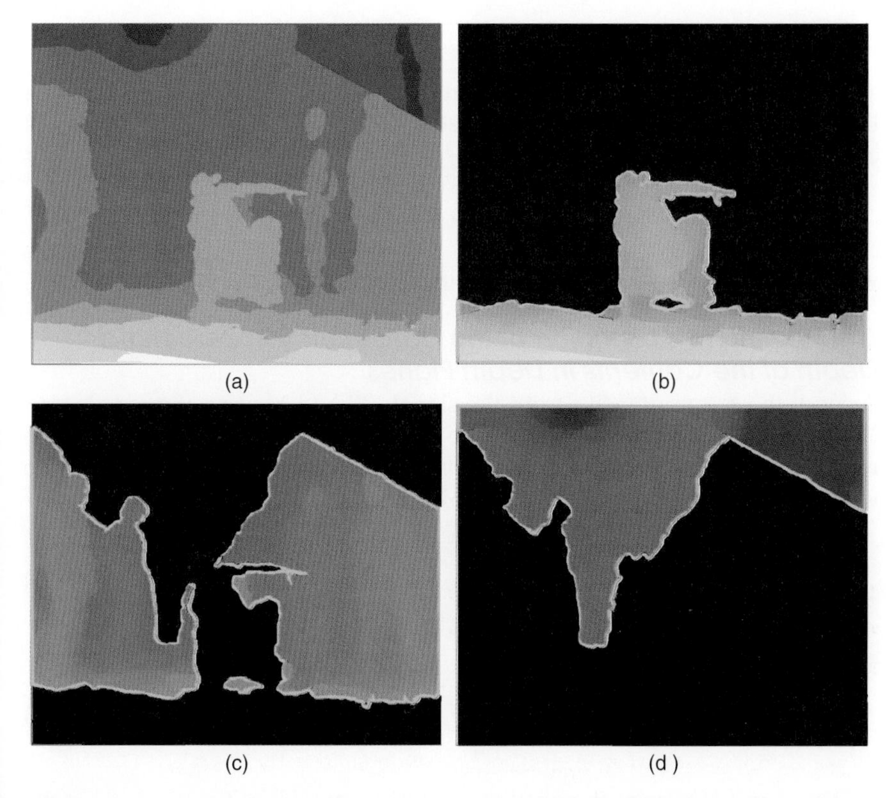

Figure 6.17 Visual depth planes of the Breakdancers sequence. (a) Original depth image (b) Segmented depth plane 1 (c) Segmented depth plane 2 (d) Segmented depth plane 3

of the depth information behind and in front of the picture respectively, relative to the screen width W. The mean intensity of the segmented depth planes is calculated for both original and degraded signals, m_x^i and m_y^j, where superscript i denotes the depth plane index. The relative distance between adjacent depth planes i and $i+1$ along the depth axis for original and the distorted signals can be identified as follows:

$$z_x^i - z_x^{i+1} = \frac{(k_{near}W + k_{far}W)}{255} \cdot (m_x^i - m_x^{i+1}) \tag{6.11}$$

$$z_y^i - z_y^{i+1} = \frac{(k_{near}W + k_{far}W)}{255} \cdot (m_y^i - m_y^{i+1}) \tag{6.12}$$

Thus, the Distortion of the Relative Distance (DRD) in depth axis for depth planes i and $i+1$, compared to the original depth map can be identified as:

$$DRD^{i,i+1} = \frac{(k_{near}W + k_{far}W)}{255} \cdot |(m_x^i - m_x^{i+1}) - (m_x^i - m_x^{i+1})| \tag{6.13}$$

$$k_{near} + k_{far} = 10 \tag{6.14}$$

and $W = 1920$ for the 42″ Philips' multi-view auto-stereoscopic display. By aggregating DRD values for all adjacent depth planes M_1 measure is evaluated.

$$M_1 = \sum_{i=1}^{n} DRD^{i,i+1} \tag{6.15}$$

where n refers to the number of visually recognized depth planes.

Calculating M_2: Distortion of the Consistency in Perceived Depth of the Contents in Depth Planes

Second, the mean intensity from the signals X^i and Y^i are removed. X^i and Y^i refer to the disparity signals of the ith depth plane. In discrete form, the resulting signals $(X^i - m_x^i)$ and $(Y^i - m_y^i)$ correspond to the projection of the vectors onto the hyperplane defined by:

$$\sum_{j=1}^{N} x_j = 0 \tag{6.16}$$

and

$$\sum_{j=1}^{N} y_j = 0 \tag{6.17}$$

Here, j and N refer to the pixel index and the total number of pixels of the corresponding depth plane. The aim of this is to make the three measurement techniques M_1 and M_2 relatively independent of each other. To model the distortion of the consistency in the perceived depth, the standard deviations of the error signal in each of the identified depth plane are evaluated. Error signal, e^i, refers to the difference of the pixel values between the processed and the original depth planes. For depth plane i the standard deviation of error can be identified as:

$$\sigma_e^i = \left[\frac{1}{N} \sum_{j=1}^{N} \left(e_j^i - \mu_e^i \right) \right]^{\frac{1}{2}} \tag{6.18}$$

where, $e^i = x^i - y^i$ and μ_e^t is the mean value of the error signal for plane i. Summation of the standard deviation of the error for all the depth planes defines M_2:

$$M_2 = \sum_{i=1}^{n} \sigma_e^i \tag{6.19}$$

Calculating M_3: Structural Comparison

Third, to evaluate M_3 after removing the mean intensity from the signals X and Y they are normalized by its own standard deviation, thus the two signals being compared have unit standard deviation. As a result, measurement techniques M_1, M_2, and M_3 are statistically independent of each other. The structure comparison is conducted on these normalized signals $\frac{(X-\mu_X)}{\sigma_X}$ and $\frac{(Y-\mu_Y)}{\sigma_Y}$. Note that the correlation between the above unit vectors is equivalent to the correlation coefficient between X and Y. In image processing, the covariance between these signals is considered as a simple but effective measure to quantify the structural similarity. Structural comparison measure in SSIM is utilized for this purpose [4]. Structural comparison is evaluated on the entire depth image (i.e. image containing all depth planes). Each image frame is divided into 16×16 macroblocks and the structural comparison (SC) is computed on macroblock basis. Structural comparison is defined as follows:

$$SC = \frac{\sigma_{xy} + k_1}{\sigma_x \cdot \sigma_y + k_1} \tag{6.20}$$

σ_x, σ_y and σ_{xy} represent the standard deviations and covariance of signals X and Y. A small constant is introduced to both numerator and denominator to avoid instability when $\sigma_x . \sigma_y$ is very close to zero.

$$k_1 = (c \cdot l)^2 \tag{6.21}$$

is chosen where c is the dynamic range of the pixel values (255 for 8-bit greyscale representation) and l as 0.001. σ_{xy} in Equation (6.20) can be estimated as:

$$\sigma_{xy} = \frac{1}{L} \sum_{j=1}^{L} (x_j - \mu_x)(y_j - \mu_y) \tag{6.22}$$

where L refers to the number of pixels of the full image. Structural error of the depth image/frame is evaluated as the average SC over the macroblocks.

$$M_3 = \frac{1}{m'} \sum_{i=1}^{m'} SC_j \tag{6.23}$$

m' and j refer to the number of macroblocks and its index respectively.

6.4.2.3 Depth Quality Model

Finally, the three components are combined to yield an overall Disparity Distortion Model (DDM):

$$DDM = f(M_1(x,y), M_2(x,y), M_3(x,y)) \tag{6.24}$$

An important point is that the three components are relatively independent. For example, the change of relative distance in depth axis and/or consistency of the perceived depth of the contents in the depth planes will not affect the structural error of the depth image. Thus the three components are combined as follows:

$$DDM = \frac{M_3(x,y)}{M_1(x,y) \cdot M_2(x,y) + k_2} \tag{6.25}$$

$k_2 = 1$ is introduced to the denominator to limit the depth quality rating between 0 and 1, and to avoid instability when $M_1.\ M_2$ is close to zero. $DDM = 1$ and $DDM = 0$ imply the maximum and minimum bounds of depth quality from the depth image respectively. This is for the reason that, at maximum depth quality $M_1.\ M_2 \rightarrow 0$ and $M_3 \rightarrow 1$ and at minimum depth quality $M_3 \rightarrow 0$.

In practice, one usually requires a single overall measure of the entire depth sequence. Thus, the mean of DDM index (MDDM) is evaluated to predict the depth quality from the depth sequence:

$$MDDM\ (X,Y) = \frac{1}{M} \sum_{j=1}^{M} DDM\ (x_i, y_i) \tag{6.26}$$

where X and Y are the reference and the distorted disparity signals respectively, x_i and y_i are the contents at the jth frame, and M is the number of frames in the depth map.

Thus, as shown in Equation (6.9) the overall depth quality from 3D video (i.e. effects from both colour texture video and depth map) can be modelled as follows:

$$Depth\_quality = [1 - VQM]^{\alpha} \cdot MDDM^{\beta} \tag{6.27}$$

Table 6.5 Performance of overall depth assessment models for 3D video sequences

Objective quality model	Depth quality		
	CC	RMSE	SSE
Average PSNR of the Rendered Left and Right views	0.7788	0.07375	0.0579
Average SSIM of the Rendered Left and Right views	0.8065	0.06745	0.05478
Average VQM of the Rendered Left and Right views	0.7753	0.07397	0.0603
Proposed Depth Quality Model	0.8716	0.03253	0.03791

Since a VQM score of 0 implies the best quality and 1 implies the worst with respect to the original video, D_M in Equation (6.9) is defined as 1-*VQM*.

Subjective ratings for depth quality from test 3 are used to assess the performance of the proposed model. Average PSNR, SSIM and VQM quality ratings of rendered left and right video are used as measures of depth quality for performance comparison of the proposed model. The relationship between the MOS for perceived depth quality and the assessment models are approximated by a symmetrical logistic function (6.3). The performance comparison metrics (CC, SSE, RMSE) for each prediction model, approximated using the symmetrical logistic function, are evaluated for all test sequences and results are presented in Table 6.5. α and β in Equation (6.9) are varied from 0 to 5 in steps of 0.5 and $\alpha = 1.5$, $\beta = 1$ revealed the best correlation with the subjective ratings. However, the optimum values of α and β will vary for different types of displays. Results imply that-the proposed model has better values for the performance comparison metrics, with regards to subjective ratings in predicting depth quality of colour-plus-depth-based 3D video. Therefore, the proposed model, based on visually important features to the brain (both monocular and binocular), can be used to predict depth quality of 3D video.

6.4.3 Compound 3D Video Quality Model

The next task is the design of a compound 3D video quality model by combining the image and depth quality models proposed earlier. In general, when different user (observer) groups and terminal characteristics are considered, the relative importance of image and depth quality of 3D video depends on content characteristics, context characteristics (display characteristics and lighting conditions) and user preference. However, in this study, effects of different context characteristics and preferences of the 3D observer are not considered. Thus, the compound 3D quality measure is modelled with respect to the content characteristics and is described as follows:

$$3D\_Quality = f_1(content) \cdot image\_quality + f_2(content) \cdot depth\_quality \quad (6.28)$$

$$f_1(content) + f_2(content) = 1 \quad (6.29)$$

6.4.3.1 Weighting Functions

Z-direction (depth direction) motion activity (ZMA) of 3D video is used to model the weighting functions f_1 and f_2. For this purpose, a depth map of the 3D video is segmented into blocks of N frames, where N refers to the frame rate of the depth map, and the standard deviation of each pixel position is evaluated for each block. After aggregating the standard deviation for all pixel positions per block the mean standard deviation over all the segmented blocks is evaluated as shown in Equations (6.30) and (6.31). This is illustrated in Figure 6.18.

$$\sigma^t_{Y_{i,j}} = \left[\frac{1}{N}\sum_{k=1}^{N}(Y^k_{i,j} - \mu^t_{Y_{i,j}})\right]^{\frac{1}{2}} \quad (6.30)$$

$$ZMA = \frac{1}{M}\sum_{t=1}^{M}\left[\sum_{j=1}^{H}\sum_{i=1}^{W}\sigma^t_{Y_{i,j}}\right] \quad (6.31)$$

$Y_{t,j}$ refers to the pixel, $\sigma^t_{Y_{t,j}}$ refers to the standard deviation and $\mu^t_{Y_{t,j}}$ refers to the mean of the pixels at i^{th}, j^{th} position of the block t. N is the number of frames per block (N= fps), k is the frame index and M refers to the number of segmented blocks of the depth map. H and W denote the frame height and width respectively. ZMA is then normalized ($nZMA$) with respect to resolution of the depth map and the dynamic range of the pixel values, i.e. $2^n - 1$ (255 for 8-bit greyscale representation), where n refers to the number of bits used for the greyscale representation of the depth map.

$$nZMA = \frac{ZMA}{(W \cdot H) \times (2^n - 1)} \quad (6.32)$$

To model weighting functions f_1 and f_2, the relative importance of the subjective ratings (MOS) for image and depth quality with respect to the subjective ratings (MOS) for overall 3D quality are analyzed, for the Interview, Orbi and Breakdancers sequences, for the quality variations considered in test 3. Here, the SSCQS method is used instead of the DSCQS method for subjective testing. This is due to the fact that, in the DSCQS method, the difference in the MOS between the undistorted original video and the distorted video, of the same sequence, is considered. Thus, the difference calculation will eliminate the effect of content characteristics of the video. Finally, f_1 and

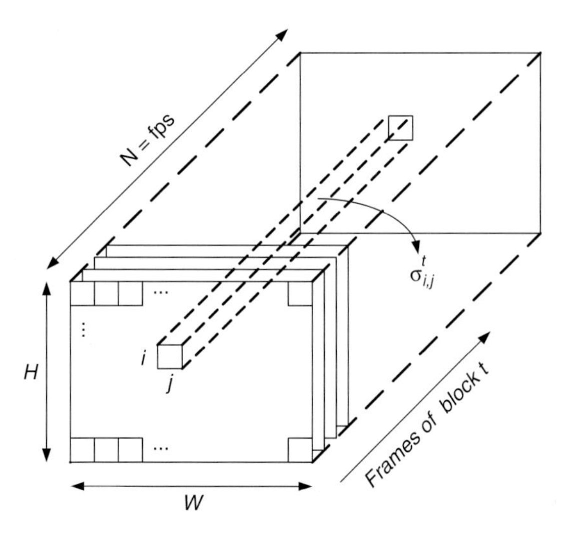

Figure 6.18 Calculation of ZMA.

f_2 are derived as functions of $nZMA$, so that they correlate with the subjectively evaluated relative importance of image and depth quality with regards to overall 3D quality, over different content characteristics. The resultant f_1 and f_2 functions are as follows:

$$f_1(nZMA) = 1 - 0.997 \cdot nZMA^{0.2393} \tag{6.33}$$

$$f_2(nZMA) = 0.997 \cdot nZMA^{0.2393} \tag{6.34}$$

The Correlation Coefficient (CC) between the objectively evaluated weighting functions and the subjectively evaluated relative importance is 0.94 (CC=1 indicates the perfect correlation). The effectiveness of these weighting functions for different content characteristics is verified for all other video sequences considered in the following section.

6.4.3.2 Compound 3D Quality Model

Incorporating the weighting functions f_1 and f_2 Equation (6.28) can be rearranged as follows:

$$\begin{aligned} 3D\_Quality = {} & (1 - 0.997 \cdot nZMA^{0.2393}) \cdot image\_quality \\ & + (0.997 \cdot nZMA^{0.2393}) \cdot depth\_quality \end{aligned} \tag{6.35}$$

By substituting the objective quality models proposed earlier for image and depth quality to Equation (6.35), the overall 3D quality model can be defined as follows:

$$3D\_Quality = (1 - 0.997 \cdot nZMA^{0.2393}) \cdot (1 - VQM_{color}) + (0.997 \cdot nZMA^{0.2393}) \cdot [(1 - VQM_{color}) \cdot MDDM_{depth\_map}] \quad (6.36)$$

Subjective tests are conducted for the quality variations considered in test 3, for 'Ballet', 'Inition 2D-3D World Cup', 'Butterfly', 'Watermill' and 'Interior' to assess the performance of the proposed compound 3D quality model. These 3D sequences contain different motion characteristics and different spatial and temporal resolutions (see Appendix A). Subjects are asked to rate the overall 3D quality using the SSCQS method. Currently, the average PSNR of rendered left and right video is widely used as a measure of overall 3D video quality. Average quality rating of rendered left and right video using PSNR and other recently developed perceptual models like SSIM and VQM are used for a more general performance comparison of the proposed model. The relationship between the MOS for overall 3D video quality and the assessment models are approximated by the symmetrical logistic function (6.3). The Performance comparison metrics (CC, SSE, RMSE) for each prediction model, approximated using the symmetrical logistic function, are evaluated for the above test sequences and the results are presented in Table 6.6.

The results clearly imply that the proposed compound 3D quality model has better values for the performance comparison metrics in predicting the overall 3D quality of colour-plus-depth-based 3D video. Therefore, the proposed model can be effectively used in predicting the perceptual quality of colour-plus-depth-based 3D video.

6.4.4 Application of the Proposed Quality Models

With the continued global growth of broadcast technologies, real-time 3D video services can be now realized. Delivery of 3DTV, movies, media,

Table 6.6 Performance comparison metrics for each prediction model

Objective quality model	Overall 3D video quality		
	CC	RMSE	SSE
Average PSNR of the Rendered Left and Right views	0.7061	0.1363	0.1091
Average SSIM of the Rendered Left and Right views	0.7387	0.0949	0.0887
Average VQM of the Rendered Left and Right views	0.8092	0.057	0.0501
Proposed Compound 3D Quality Model	**0.8461**	0.0337	0.0319

advertising or other popular 3D content to a set-top box through a broadcast system such as broadband, wireless or satellite network connection helps service providers round out their 'triple-play' service offering, and to remain competitive in a rapidly evolving marketplace. The acceptance and success of a 3D broadcast system depend on how well the system performs. It is very important to optimize the end-to-end 3D video chain to enable cost-efficient delivery of rich 3D media services in real-life environments by jointly addressing user experience and quality/resource trade-offs. The quality of the perceptual attributes of 3D video is one of the evaluation criteria to assess the performance of a 3D broadcast system. Hence, the systems' parameters can be tuned and optimized to the customers' quality preference.

Due to the bandwidth restrictions, it is necessary to preprocess and compress 3D content before transmission. This is equally applicable to broadcasting services as well as real-time streaming scenarios such as IPTV. In such situations pre-processing and compression need to be carried out in a manner so as not to compromise the perceived quality attributes of 3D video. Coding decisions can effectively be taken using the proposed quality models in the compression life cycle, replacing simple statistical metrics such as PSNR, MSE or SAD (Sum of Absolute Difference) that do not consider 3D perception of humans. Further, in online streaming applications, delay, jitter and packet loss may cause unacceptable 3D viewing conditions. A reduced reference version of the proposed quality models can be used at network nodes to predict the effects of network conditions on perceptual attributes of 3D video and to improve the perceived quality of the received data by error concealment.

Availability of such objective quality models would allow a 3D video service infrastructure that focuses on the user's experienced 3D quality and presents a seamless, trusted and, most importantly, a satisfying experience to benefit both service provider and end user.

6.4.5 Context Dependency of Visual Experience

Visual experience is context dependent. Display technologies, ambient properties, communication channel capabilities and conditions, users and content types are some of the contextual factors that determine the image quality and depth perception. For example, when 3D video is watched on a traditional 2D television using multichrome filters (i.e. anaglyph glasses), each eye sees distorted colours and colour perception relies on the brain's ability to infer the colours by combining both images and the users' pre-existing experience of real-world colours.

On top of image quality and depth perception, there are other aspects of usability and comfort that are not strictly related to the previous qualities, but still affect the experience of the user. User experience depends on the correctness of the displayed image as well as technical parameters like field of view, motion parallax continuity, maximum visible depth, which all affect usability in some way.

3D reproduction quality can be expressed more in terms of lack of error rather than quality measurements. Various 3D display technologies lack 3D reproduction from certain angles, or certain aspects of 3D vision altogether, such as horizontal and/or vertical motion parallax, depth perception from the whole field of view (FOV), or acceptable 3D view distance. Correctness of 3D image reproduced by the display is also an important factor. Some displays show 3D images that are not physically consistent (e.g. the images have bad perspectives) since they flatten the image to keep it inside the visible depth range.

6.4.6 3D-Specific Technical Properties that Affect the Viewing Experience

6.4.6.1 View Uniformity

The uniformity of multi-channel systems largely depends on any variation in brightness, contrast and colour balance between the neighbouring views of the display.

6.4.6.2 Total Field of View (FOV)

Auto-stereoscopic 3D displays provide areas of different size from which the image can be perceived. Typical values range from 10 to 60 degrees total field of view, depending on the definition (i.e. where the viewer stands with respect to the centre of the screen). This measure does not apply to wearable visual aids (such as shutter glasses, polarized glasses, etc.) displays, as the viewer can see the same image regardless of his/her position. Whether the viewer can see a correct 3D image from everywhere inside this range or just a specific sweet spot also affects the quality of experience. FOV can be measured by finding the range from which the full 3D image can still be correctly seen on both sides.

6.4.6.3 Angular Resolution and Field of Depth (FOD)

In the case of 3D displays, there is an additional measure of resolution to take into account (on top of the usual horizontal and vertical). The number of independent beams emitted inside the FOV determine the angular resolution of the display. This property shows us the 'jumpiness' of the 3D display. Moreover, this same value directly relates to the Field of Depth, which is the range of depth that the display is able to show without excessive blurring. Angular resolution can be measured with a specialized measuring device such as an Eldim VCMaster3D. Typical Field of View for 3D display technologies is shown in Figure 6.19.

6.4.6.4 Cross-Talk

Cross-talk defines how much a neighbouring view affects the currently measured view. Any cross-talk between views is measured by illuminating

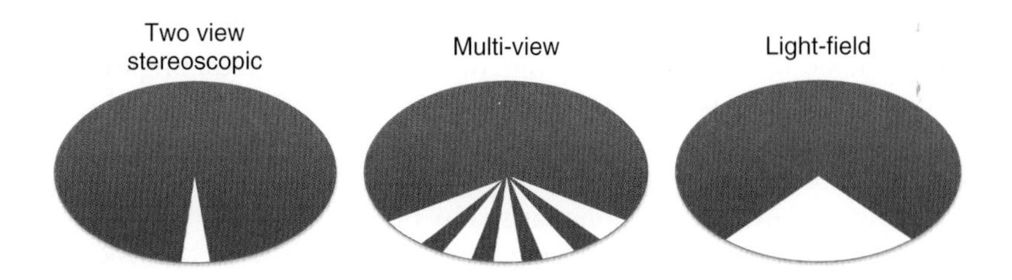

Figure 6.19 Typical Field of View for 3D display technologies

a single view of the display using a mid-grey background while measuring the brightness of the screen at its neighbouring views. This and the view uniformity can also be measured using Eldim EZLite3D.

6.4.6.5 Comfort

Some users may not be happy to wear filter glasses. Polarization-based techniques also require glasses. Depending on type of polarization, this technique may also prevent the user from tilting his/her head. Although shutter glasses-based 3D displays provide a full resolution 3D image, which is considered important, they usually cause headaches for most users, especially if the refresh rate is not sufficiently high. This severely affects the users' comfort, even when the image quality and depth perception are sufficiently good. The auto-stereoscopic technique reproduces colour images on both eyes without any support from wearable visual aids. However, correct 3D vision in most auto-stereoscopic displays is limited to special angles, which are identified as 'sweet spots'. Even if the viewer can see the correct image by holding his/her head strictly in the sweet spot, he/she may feel uncomfortable after a while. Therefore, such restrictions may prevent the viewer from enjoying 3D visual contents. However, modelling this is extremely difficult without proper subjective tests.

6.5 Conclusion

3D video in the format of monocular video (colour texture video) augmented by the greyscale depth map is a promising representation technique, which provides high quality 3D video with a smaller bandwidth. This chapter has addressed the possibility of modelling important attributes of the above format of 3D video, which could accelerate the further development of broadcast technology and the introduction of new 3D video services in time. First, subjective results are analysed to investigate the effects of monocular colour texture video and the depth map on image and depth quality attributes as experienced by the viewers. Second, the possibilities of modelling these attributes are investigated. To predict the image quality of 3D video, three widely used 2D quality models have been considered with four different combination approaches incorporating different aspects of visual characteristics of the observers. The quality rating of the colour texture video evaluated using VQM showed better potential in predicting the image quality of 3D video. To model depth quality of 3D video, a novel quality assessment framework, which could estimate separately the monoscopic and stereoscopic contributions, is presented. While the former assesses the trivial monoscopic perceived distortions caused by blur, noise and contrast change, the latter assesses the perceived degradation of visually recognized depth planes of the 3D content. Due to the lack of suitability of existing 2D quality models to evaluate the depth map for its contribution to binocular cues, the MDDM metric is proposed by combining visually important features to the brain. Results imply that the proposed model offers a better performance in predicting the depth quality of 3D video compared to existing methodologies.

Developing a single quality metric for 3D video still remains a big challenge due to the number of perceptual attributes, such as image quality, depth quality and visual comfort, associated with the 3D experience, and their relative importance to the brain. As a simplified solution to the above problem, a compound 3D quality model is designed by combining the dominant perceptual attributes of 3D video, i.e. image and depth quality. Subjective tests are conducted to verify the performance of the proposed model and results clearly imply that the proposed compound 3D quality model has better performance in predicting overall quality of 3D video compared to existing methodologies. Moreover, the performance of the proposed model is further justified for transmission errors of an IP network. This implies that, while subjective test results remain as the best and precise judgement of 3D video quality, the use of proposed quality models is an acceptable compromise for the 3D video research community to speed up the development of 3D consumer products, services and consumer products.

References

[1] Bech, S., Hamberg, R., Nijenhuis, M., Teunissen, C., De Jong, H.L., Houben, P. and Pramanik, S.K. (1996) 'The RaPID perceptual image description method (RaPID)', *Proceedings of SPIE*, **2657**, 317–328.

[2] Branden Lambrecht, C.J. and Verscheure, O. (1996) 'Perceptual quality measure using a spatio-temporal model of the human visual system', *Proceedings of SPIE*, **2668**, 450–461.

[3] Winkler, S. (2005) *Digital Video Quality: Vision Models and Metrics*, John Wiley & Sons, Ltd, Chichester.

[4] Wang, Z., Lu, L. and Bovic, A.C. (2004) 'Video quality assessment using structural distortion measurement', *Signal Processing: Image Communication*, **19** (2), 121–132.

[5] Pinson, M.H. and Wolf, S. (2004) 'A new standardized method for objectively measuring video quality', *IEEE Transactions on Broadcasting*, **50** (3), 312–322.

[6] Video Quality Experts Group (VQEG) (2003) 'Final report from the video quality experts group on the validation of objective models of video quality assessment, phase II', Tech. Rep., VQEG.
[7] Julesz, B. (2006) *Foundations of Cyclopean Perception*, MIT Press, Cambridge, MA.
[8] Baecke, S., Lützkendorf, R., Hollmann, M., Macholl, S., Mönch, T., Mulla-Osman, S. and Bernarding, J. (2006) 'Neuronal activation of 3D perception monitored with functional magnetic resonance imaging', in Deutsche Gesellschaft für Medizinische Informatik, Biometrie und Epidemiologie e.V. (GMDS). **51**, Jahrestagung der Deutschen Gesellschaft für Medizinische Informatik, Biometrie und Epidemiologie. Leipzig, 10.–14.09.2006. Düsseldorf, German Medical Science, Cologne.
[9] Vanduffel, W., Fize, D., Peuskens, H., Denys, K., Sunaert, S., Todd, J.T. and Orban, G.A. (2002) 'Extracting 3D from motion: Differences in human and monkey intraparietal cortex', *Science*, **298** (5592), 413–415.
[10] International Telecommunications Union (ITU) Radio Communications Sector (2000) 'Subjective assessment of stereoscopic television pictures', ITU BT.1438, pp. 14.
[11] International Telecommunication Union (ITU) Radio Communication Sector (2002) 'Methodology for the subjective assessment of the quality of television pictures', ITU-R BT.500-11, p. 48, January.
[12] Aksay, A., Bilen, C., and Akar, G.B. (2005) 'Subjective evaluation of effects of spectral and spatial redundancy reduction on stereo images', in *13th European Signal Processing Conference*, Antalya, Turkey, September.
[13] Seuntiëns, P., Meesters, L. and IJsselsteijn, W. (2003) 'Perceptual evaluation of JPEG coded stereoscopic images', *Proceedings of SPIE*, **5006**, 215–226.
[14] Wang, Z. and Simoncelli, E.P. (2005) 'Reduced-reference image quality assessment using a wavelet-domain natural image statistic model', in *IS&T/SPIE's 17th Annual Symposium on Electronic Imaging*, San Jose, CA, January.
[15] International Telecommunication Union (ITU) Telecommunication Sector (1999) 'Subjective video quality assessment methods for multimedia applications', ITU-T P.910, Series P, pp. 37, September.
[16] Wandell, B.A. (1995) Foundations of Vision, Sinauer Associates, USA, May.
[17] Hekstra, A.P., Beerends, J.G., Ledermann, D., de Caluwe, F.E., Kohler, S., Koenen, R.H., Rihs, S. Ehrsam, M. and Schlauss, D. (2002) 'PVQM: A perceptual video quality measure', *Signal Processing: Image Communication*, **17** (10); 781–798.
[18] Boev, A., Gotchev, A., Egiazarian, K., Aksay, A. and Akar, G.B. (2006) 'Towards compound stereo-video quality metric: a specific encoder-based framework', in *Proceedings of IEEE Southwest Symposium on Image Analysis and Interpretation*, **2006**, 218–222.
[19] Furht, B. and Marqure, O. (2003) *The Handbook of Video Databases: Design and Applications*, CRC Press, Boca Raton, FL.
[20] ISO/IEC JTC 1/SC 29/WG 11 (2006) 'Committee draft of ISO/IEC 23002-3 auxiliary video data representations', in WG 11 Doc. N8038, Montreux, April.

7

Conclusions and the Future of 3DTV

The recent renaissance in 3D entertainment has been made possible by advances in digital image processing technology. Such advances have made it possible to precisely calibrate display systems, perform post-processing on captured images, and compress video to a level suitable for distribution to consumers. The impact of these advances is that it is now easier to provide a more consistent and more comfortable experience to the viewer than has previously been seen.

Many television sets are now being sold that are capable of displaying stereoscopic content with the aid of synchronized shutter glasses. This development has been accompanied by the increasing availability of 3DTV broadcast channels, and by 3D Blu-Ray players. The amount of available 3D content has also grown significantly, with movies continuing to be made in 3D, or converted to 3D in post-production.

All of these topics have been discussed in this book. This chapter summarizes the content of the chapters, and also looks ahead to what developments are required to make 3DTV a lasting success.

7.1 Chapter Summary

7.1.1 Chapter 1: Introduction

The first chapter looks at the history of 3D images and video. 3D has been around in some form for more than 100 years. During that time its popularity has risen and fallen. Historically, a major issue has been that it is very difficult to deliver a consistent, high quality 3D viewing experience. In the

3DTV: Processing and Transmission of 3D Video Signals, First Edition.
Anil Fernando, Stewart T. Worrall and Erhan Ekmekcioğlu.

1950s boom, many movie theatres were found to be presenting content that caused significant visual discomfort to viewers, and even induced feelings of nausea. Digital technology has enabled stereoscopic video to be post-processed, and displays to be carefully calibrated. This means that it is now possible to provide a guaranteed minimum Quality of Experience to viewers.

The first chapter also examines different 3D video formats, including different stereoscopic frame packing formats, and the multi-view plus depth format. The inclusion of depth information makes it easier for displays to render intermediate views in between the real captured viewpoints. This facilitates applications such as free-viewpoint video, where the user can control their viewing angle interactively.

7.1.2 Chapter 2: Capture and Processing

The second chapter of the book deals with the capturing and preparation of the content, which is one of the most important elements in the 3DTV chain. The critical problems associated with the early 3D content/movie productions included vertical parallaxes between the left eye and the right eye channels, resulting from inaccurate rig construction; imperfections in inter-axial distance computation disregarding perceptual comfort over long periods of viewing; and also inconsistencies in processing of the left eye and the right eye channels after filming.

All of these shortcomings have led to the careful and meticulous consideration and set-up of 3D media acquisition and post-processing mechanisms. The cameras need to be physically positioned in a way to prevent discomfort for viewers. The cameras must also be carefully configured so that their settings match precisely (e.g. focal length, shutter speed, etc.) for optimum perceptual quality free of visual degradations.

The introduction of digital technology allowed a greater tolerance of imperfections in the physical camera configuration. Digital post-processing techniques allowed the camera views to be aligned by rectification, and balanced the colour differences between them. This chapter explores how stereoscopic and multi-view video can be captured using specially configured camera rigs. In addition, it looks at the generic post-processing techniques. Depth extraction techniques are outlined, which are essential for generating content in the multi-view plus depth (MVD) format.

7.1.3 Chapter 3: Compression

Compression is a substantial stage in the end-to-end 3DTV chain, where a huge load of raw video information needs to be carried over bandwidth-constrained networks. Compression had always played an important role in traditional video communication systems, by enabling the video information to be represented by a smaller number of bytes required to represent the original signal, without any loss of information. Usually the volume of visual data in 3D video communications is considerably larger than 2D video, depending on the number of camera viewpoints that are delivered to the end users. This could be two views only (e.g. stereoscopic 3D video) for a standard 3DTV, or over a 100 views to drive possibly a light-field or holographic display.

The traditional video coding approaches applied to 2D video signals are quite applicable to 3D video representation formats, despite their inability to remove the majority of inherent spatial redundancies in them. Thus, more sophisticated compression algorithms are needed for 3D video coding in order to extract the majority of inter-camera statistical correspondences and reduce the bit-rate of the resultant compressed representation. This chapter outlines the details of widely deployed modern 3D and multi-view video compression techniques. Video coding principles are briefly provided that are followed by the explanation of standards on 3D video coding. Finally, this chapter provides detailed explanation on the most used forms of 3D video, i.e. stereoscopic, multi-view and multi-view with depth map (MVD).

7.1.4 Chapter 4: Transmission

The renaissance of 3D also coincides with a revolution in the way that content is delivered to and consumed by the public. In addition to traditional broadcast delivery systems, it is also necessary to deliver content over the Internet to people's homes, and also to their mobile devices. Adaptive Bit-Rate (ABR) services have seen significant growth in recent years, which will be further bolstered by the introduction of the MPEG-DASH standard. Initially, 3D services have been offered by packing two stereoscopic frames into a single video frame, making 3D transparent to the whole delivery system. More advanced 3D content presents a different delivery challenge to traditional 2D content. The presence of depth maps and multiple views represents a complication that stretches further than a simple increase in bit-rate. Each elementary stream (e.g. depth or individual view bitstreams) may have a different importance in terms of reconstructing the 3D video, and in terms of importance for end-user Quality of Experience (QoE). There are significant opportunities to exploit these issues to optimise the end-user QoE. However, additional research is still needed on the effects of losses on perceived quality and viewer fatigue. The latter is likely to be very important, as some error resilience schemes may provide short term gains in perceived QoE, but at the expense of longer term eye strain and fatigue.

7.1.5 Chapter 5: Rendering and 3D Displays

3D video can be visualized in a variety of ways, in contrast to the traditional 2D media. This can range from the widespread available plano-stereoscopic 3DTV, where the illusion of depth is present without the possibility of changing the viewing angle, to more realistic 3D screens, where the source of light emitted from the virtual objects is replicated. A number of 3D display technologies are available, all of which differ in terms of applied image processing and rendering techniques and the requirements for input media size and format. Thus, the mode of displaying the reconstructed 3D scene also show differences. A standard plano-stereoscopic 3D display on the market can display only left-eye and right-eye channels, regardless of the viewing angle, whereas multi-view displays usually have the capability of displaying more than two views simultaneously, so that different viewers looking from different angles would see different pictures. A series of dedicated processing tasks within the display are adopted for the 3D mode of viewing. This chapter explains the stages of 3D video rendering and outlines the details of various 3D display technologies. In addition, this chapter also outlines several elements of various adaptation schemes devised for 3D media systems, while outlining the inherent differences from the 2D media adaptation.

7.1.6 Chapter 6: Quality Assessment

The quality assessment chapter looks at factors affecting the human perception of 3D images and video. There are a significant number of cues used by the HVS to interpret depth. Stereoscopic disparity is just one such cue. Many cues can be considered monoscopic (i.e. can be perceived using only a single view). The chapter also reviews existing quality assessment techniques, and describes in detail an effective approach to the modelling of human perception of 3D video quality. The described quality assessment technique is shown to correlate closely with expert viewer assessment of 3D video quality.

7.2 The Future of 3DTV

Despite these advances, the future of 3DTV cannot be considered to be secure unless advances continue to be made in a number of areas. This chapter outlines the areas where progress is needed and discusses their importance with regards to the continuing spread of 3DTV.

- Understanding of human 3D perception
- Display technologies
- Production approaches and technologies
- Compression algorithms

7.2.1 Understanding of Human 3D Perception

Some of the issues that need to be solved are directly related to technology. However, to fully optimize 3D technology it is necessary to gain a better understanding of the impact of 3DTV on the Human Visual System (HVS). Some work has already been carried out to establish models that relate compressed video quality to the Quality of Experience (see Chapter 6). Additional work on topics such as Just Noticeable Difference (JND) is also helpful in aiding video compression algorithms. JND can help a compression algorithm retain perceptually important depth differences.

All of the above are important, and further work on these topics is needed to help improve 3D video quality. However, probably the most important factor in ensuring the success of 3D video is understanding visual comfort and fatigue. There is already a considerable body of knowledge concerning issues such as accommodation-convergence mismatch (see Section 6.2). What is missing are reliable methods to automatically take into account such issues when creating 3D content, and when displaying 3D content. This is a particular problem for 3DTV, because an important factor in ensuring visual comfort is making sure that the disparity between two stereoscopic images is not too large. Whether the disparity is too large depends upon the screen size, and the distance from the screen. Therefore, calibration is required to match a screen to its viewing environment.

Cinemas are a controlled environment, which can be carefully calibrated to ensure visual comfort.[1] For the home, calibration should take place with as minimum user input as possible. Ideally, a 3DTV should detect viewer positions and change the disparity between the two views so that all viewers can watch comfortably. The major issue here is to try and take knowledge of visual comfort and produce a mathematical model that can be used by a 3DTV. Such a mathematical model can also be used in content production tools, so that content creators can understand the type of viewing conditions in which their raw captured content is suitable. In post-production, content creators can make adjustments to their content to optimize it for a particular environment (e.g. cinema or home use).

7.2.2 Display Technologies

The most widespread technologies rely on wearing glasses. There are a number of issues with glasses-based technology:

- Both polarizing and shutter glasses reduce the amount of light reaching the eye.
- A significant number of people do not like to wear 3D glasses, due to physical discomfort or for aesthetic reasons.

3D glasses are perhaps more of a problem for 3DTV than cinema, when one considers the differences between how the content is consumed. Television

[1] Although cinemas can be carefully calibrated, the wide range of viewing positions means that they are not necessarily ideal environments for comfortable viewing of 3D video.

consumption may often be done in the presence of friends and/or family, where there may be social interaction. In this context, viewers may be more concerned about wearing "silly" glasses, than in a dark cinema where there is little social interaction.

All of these factors mean that there is a strong need for glasses-free 3D displays. This is a major concern for 3DTV, because there are no glasses-free technologies on the horizon, which are both of high quality and affordable for the consumer (see Chapter 5 for an overview of glasses-free technologies). The two technologies most likely to be exploited in consumer devices are:

- *Parallax barrier displays* – these displays are relatively cheap, but only stereoscopic content can be viewed within a narrow viewing angle. Outside of this viewing angle the picture cannot be seen correctly. This makes the technology suitable only for one viewer, and is therefore inadequate for 3DTV.
- *Lenticular displays* – these displays cost significantly more to manufacture, compared to shutter glass and parallax barrier-based displays. However, they are capable of offering multiple "sweet spots", which allows multiple viewers to experience 3D. One issue is that they work best when provided with multiple views (i.e. more than two). This makes it more difficult to produce suitable content, and also to deliver it to the home over the existing broadcast or IPTV infrastructure.

Of these two technologies, lenticular displays are the most suitable. However, adopting them will incur considerable expense for consumers, content creators, and broadcasters. Content creators and broadcasters are affected by the need to provide video with additional views. Therefore, it seems likely that most 3DTVs will employ shutter glass technology, which has the key advantage of being relatively cheap, and simple to integrate without fundamentally changing the display technology. This means that there will be a risk that 3DTV will fail, because consumers tire of having to wear 3D glasses.

7.2.3 Production Approaches and Technologies

There has been a significant increase in the availability of 3D content. This has been made possible by the introduction of 3D Blu-Ray, and by broadcasters introducing 3D television channels. Clearly, content creators are already finding it possible to produce stereoscopic content. However, there are two key benefits that may be derived from further improving production technologies and workflows:

- Maximizing quality for consumers to ensure high satisfaction levels, and to accelerate the take-up of 3DTV.

- Reducing the time and costs associated with creating 3D content, so that more content is made available.

Two key areas are discussed in the following sub-sections: 3D video processing techniques, and 3D video capture. Both of these pose significant challenges for content creators, and will therefore benefit from improvements in approaches and technologies.

7.2.3.1 3D Video Capture

3D video capture is made highly challenging by the current requirement to use multiple cameras attached to a special rig. The video capture system needs to be carefully adjusted and calibrated to ensure that views are captured with the optimum disparity. The situation is challenging with two cameras, where simple issues such as zooming[2] can cause difficulties. As more cameras are added, the challenges become greater. System configuration and calibration become even more time-consuming. Currently, it seems impractical to film with more than four high quality cameras.

The size of 3D video capture equipment is also potentially an issue. Current approaches involve putting two high quality cameras on a single rig. Because high quality cameras are generally very large, the disparity between them would often be too large to be used for stereoscopic content capture. Therefore, it is often necessary to use a mirror system in conjunction with the cameras. The mirror system reflects light into the cameras from two viewpoints with a narrow enough disparity for use in stereoscopy. All of these factors mean that high-end systems are large and unwieldy. Manufacturers are just starting to produce high-end cameras with two integrated lenses for stereoscopic content capture. This will make 3D capture less unwieldy, and will allow 3D content creation in a wider range of scenarios.

While work is going on to make stereoscopic content capture more simple, multi-view content capture will remain very challenging for some time. One method of providing multi-view content, without having to use large numbers of cameras, is to exploit Time-of-Flight (ToF) camera technology. These types of cameras provide depth information, which can be used to render additional views in post-production. The current problems with such cameras are cost, resolution, and accuracy of the depth information that they generate. There are also potential issues with ensuring that the depth information is correctly aligned with the colour images captured by other cameras on the same rig.

[2] Zooming effectively changes the disparity between two viewpoints.

7.2.3.2 3D Processing Techniques

Even if content is captured carefully, with precisely calibrated systems, the 3D video will not be in a suitable format for use in every 3DTV system, or in a cinema. There are a number of issues that make 3D video processing techniques vital to easing 3D content creation workflows:

- The many different formats that are required for different 3D systems (see Section 1.2).
- The viewing arrangements for particular content (e.g. how the video must change to match the required disparity to the viewing environment, and viewer seating positions).
- Incorrect calibration and setting-up of cameras during filming. For example, if the cameras are slightly misaligned, then viewers would experience significant visual discomfort. Correcting these misalignments is vital.
- Sometimes camera calibration and alignment data can only be captured "through the lens", which requires image processing techniques to estimate relative alignment, and camera properties.

Clearly, image processing technologies are required to convert images between formats (e.g. synthesis of depth maps, multi-view to stereoscopic conversion). Such techniques already exist, but techniques such as depth map extraction cannot be said to be mature yet. Even small misalignments between colour and depth information can result in significant visual artefacts. Therefore, continued research on developing high quality algorithms is needed.

Disparity warping algorithms have already been demonstrated. The key remaining challenge is to be able to implement real-time, high quality, low latency view warping algorithms in consumer products. Such view warping will be needed to calibrate 3DTVs precisely to different home environments.

The remaining topics (realignment of views, calibration) are also areas where algorithms have already been developed. As for disparity warping, the key challenges are improving the quality, while reducing the complexity of the algorithms.

7.2.4 Compression Algorithms

Compression efficiency is a very important issue, as broadcasting additional views can increase costs significantly for broadcasters and make 3D services expensive. If multi-view displays are to be supported, then better compression efficiency is vital.

A significant amount of work is going on in the MPEG 3DV group to improve the performance of 3D compression algorithms. Standardization of a new video codec, called HEVC, will be finalized in 2012. A new 3D extension to HEVC will be finalized in 2013–2014. As HEVC has a much better performance as regards compression than H.264, its 3D extension is likely to perform much better than MVC, which uses H.264-based compression algorithms.

Although compression efficiency is important, the complexity of the compression algorithm is also critical. The complexity will impact on the cost of the consumer equipment required to decompress the 3D content. One aspect that can significantly impact on the ease of implementation is how correlation between views is exploited. Exploiting the correlation is important to improve the compression efficiency. However, creating dependencies between viewpoints also limits opportunities for compressing and decompressing the multiple views in parallel. This is becoming an important issue, as many real-world systems are increasingly exploiting parallel processing opportunities. Multi-core processing is widely seen as an important factor in improving processing speeds.

7.2.5 Looking Further Ahead

Of all of the many challenges that must be solved to maintain the popularity of 3DTV, the display technology is probably the most vital. The display technology has a major impact on the end users. Although some of the more exotic technologies (e.g. light-field and holographic displays) are not yet ready for mainstream use, they may prove to be an indication of what might be seen in people's homes in the longer term. Displays that produce an image that has real physical depth are important in removing the accommodation-convergence visual discomfort issue. They can also provide a real 3D experience without glasses or sweet spots.

Aside from the technological challenges involved in developing the displays, there are probably two further major challenges that must be overcome to support these displays: content creation, and display aesthetics. The last problem might sound trivial, but the appearance of devices plays an important role in their uptake. Displays that produce light with physical depth usually require a box with physical depth. Producing an aesthetically pleasing flat panel version would be very challenging. The issue of content creation cannot be dismissed, from a technical point of view (large numbers of views are required), or from an artistic point of view. Content creators are currently used to producing visual content where they control precisely what the viewer sees. Allowing movie viewers to move around a scene will require rethinking how to present the story.

Appendix A

Test Video Sequences

A.1 2D Video Test Sequences

Some basic details of the 2D video sequences used for simulations in this book are given below.

Table A.1

Frame #0	Sequence name	Frame format	Temporal resolution (fps)	Motion activity level
	Claire	QCIF (spatial res: 176 × 144); sampling: 4:2:0	30	Low
	Foreman	QCIF (spatial res: 176 × 144); sampling: 4:2:0	30	Medium
	Soccer	QCIF (spatial res: 176 × 144); sampling: 4:2:0	30	High

3DTV: Processing and Transmission of 3D Video Signals, First Edition.
Anil Fernando, Stewart T. Worrall and Erhan Ekmekcioğlu.

A.2 3D Test Video Sequences

Some basic details of colour-plus-depth-based 3D video sequences used for the subjective tests in the book are given below.

Table A.2

Frame #0	Sequence name	Spatial resolution	Temporal resolution (fps)	Motion activity level
	*Orbi*	720 × 576	25	Medium/high (parallel camera motion)
	*Interview*	720 × 576	25	Low (static camera)
	*Breakdancers*	720 × 576	15	High (static camera)
	*Ballet*	720 × 576	15	Medium (static camera)
	*Inition 2D-3D World Cup*	960 × 540	25	High (parallel camera motion)
	*Butterfly*	960 × 540	25	Low (static camera)
	*Watermill*	960 × 540	30	Medium/High (parallel camera motion)
	*Interior*	960 × 540	25	Medium (parallel camera motion)

Table A.2 *(continued)*

Frame #0	Sequence name	Spatial resolution	Temporal resolution (fps)	Motion activity level
	T-Rex (computer generated)	1024 × 768	25	Medium (4 views static camera)
	Pantomime	1280 × 960	30	Medium (80 views static camera)
	Cafe	1920 × 1080	30	Low (9 views static camera)

Appendix B

Introduction to the Experiment and Questionnaire

The handout given to subjects, before the test, explaining the procedure of the experiment is given in Section B.1. The questionnaire given to the subjects on the day of the experiment is given in Section B.2.

B.1 Introduction to the Experiment

Thank you for participating in this experiment.

In this experiment, you are asked to assess some 3D video sequences in terms of an evaluation attribute (i.e. Image quality or depth quality or overall 3D quality), which will be clarified for you. The sequences will be displayed as shown in Figure B.1.[1] The images/sequences are displayed on a 3D screen. Prior to each video sequence, a grey screen is displayed for 3 seconds. Each video sequence will be displayed for 10 seconds.

Video sequences A and B will be displayed on the screen consecutively. Then the same sequences will be repeated again. With the start of the repetition, you can start ranking individual sequences for their perceived quality attribute. In order to complete the assessment process a grey scale will be displayed for another 5–11 seconds after repeating the video sequences.

[1] When the SSCQS method is used instead of the DSCQS method, the handout is changed accordingly.

3DTV: Processing and Transmission of 3D Video Signals, First Edition.
Anil Fernando, Stewart T. Worrall and Erhan Ekmekcioğlu.

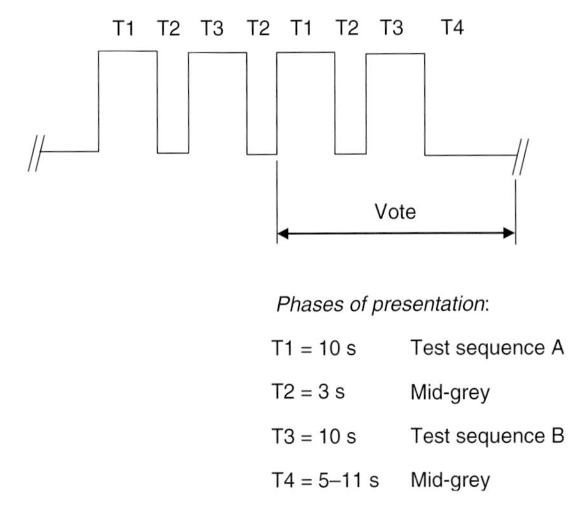

Figure B.1 Presentation structure of video sequences during the assessment

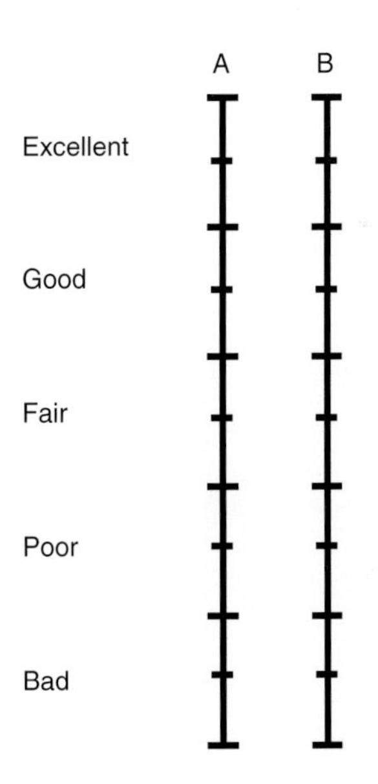

Figure B.2 Assessment scale (DSCQS scale)

The assessment will take place on a paper provided to you on the evaluation day. The scale given to mark your scores is shown in Figure B.2. Assess all the sequences in terms of their perceptual quality attribute (i.e. Image quality or depth quality or overall 3D quality). You are expected to use the full range of the scale. The labels above the scale serve as a reference, i.e. the bottom of the scale is the lowest and the top is the highest score possible.

Before the experiment begins a short training session will take place. Through this training you will get acquainted with the videos that are used in the experiment.

It is important to watch the whole video, because the differences are not the same for every video. You are allowed to move the position of your head only by a centimetre to the left and to right to get a better view of the image. If you have any questions, please ask them during the training session. After the training session the experiment begins. The experiment consists of two 15-minute sessions. In between there will be a short break of 10 minutes.

Again, thank you for participating in this experiment.

B.2 Questionnaire

Name of participant:

Age:

Test number:

Have you participated in subjective tests before?
(Yes/No, if yes, please specify type of test, i.e. 3D or 2D video):

Have you experienced any 3D video content before? (Yes/No):

Symptom Check

1. Before the test

- **General discomfort** a) none b) slight c) moderate d) severe
- **Fatigue** a) none b) slight c) moderate d) severe
- **Headache** a) none b) slight c) moderate d) severe
- **Difficulty focusing** a) none b) slight c) moderate d) severe
- **Blurred vision** a) none b) slight c) moderate d) severe

2. After the test

- **General discomfort** a) none b) slight c) moderate d) severe
- **Fatigue** a) none b) slight c) moderate d) severe
- **Headache** a) none b) slight c) moderate d) severe
- **Difficulty focusing** a) none b) slight c) moderate d) severe
- **Blurred vision** a) none b) slight c) moderate d) severe

Comments

....

....

....

....

....

Office use only

Vision characteristics of the participant

1. Eye dominance: Left Right

2. Left eye acuity: Right eye acuity:

3. Colour blindness: Yes No

4. Stereoscopic vision: Yes No

5. Stereoscopic sensitivity:

Date:

Name of experimenter:

Index

3DTV: Processing and Transmission of 3D Video Signals, First Edition.
Anil Fernando, Stewart T. Worrall and Erhan Ekmekcioğlu.